Paraffin - an Overview

Edited by Fathi Samir Soliman

Published in London, United Kingdom

IntechOpen

Supporting open minds since 2005

Paraffin - an Overview
http://dx.doi.org/10.5772/intechopen.75234
Edited by Fathi Samir Soliman

Contributors
Keshawa Shukla, Mayank Vishal Labh, Amir Reza Vakhshouri, Erika C.A. Nunes Chrisman, Márcio N. Souza, Angela C. P. Duncke, Marcia C.K. Oliveira, Wenye Lin, Zhenjun Ma, Haoshan Ren, Jingjing Liu, Kehua Li, Fathi Samir Soliman

Notice
Statements and opinions expressed in the chapters are these of the individual contributors and not necessarily those of the editors or publisher. No responsibility is accepted for the accuracy of information contained in the published chapters. The publisher assumes no responsibility for any damage or injury to persons or property arising out of the use of any materials, instructions, methods or ideas contained in the book.

First published in London, United Kingdom, 2020 by IntechOpen
IntechOpen is the global imprint of INTECHOPEN LIMITED, registered in England and Wales, registration number: 11086078, 7th floor, 10 Lower Thames Street, London, EC3R 6AF, United Kingdom
Printed in Croatia

British Library Cataloguing-in-Publication Data
A catalogue record for this book is available from the British Library

Additional hard and PDF copies can be obtained from orders@intechopen.com

Paraffin - an Overview
Edited by Fathi Samir Soliman
p. cm.
Print ISBN 978-1-83880-594-4
Online ISBN 978-1-83880-595-1
eBook (PDF) ISBN 978-1-83880-596-8

We are IntechOpen,
the world's leading publisher of
Open Access books
Built by scientists, for scientists

5,000+
Open access books available

125,000+
International authors and editors

140M+
Downloads

151
Countries delivered to

Our authors are among the

Top 1%
most cited scientists

12.2%
Contributors from top 500 universities

CLARIVATE ANALYTICS
BOOK
CITATION
INDEX
INDEXED

WEB OF SCIENCE™

Selection of our books indexed in the Book Citation Index
in Web of Science™ Core Collection (BKCI)

Interested in publishing with us?
Contact book.department@intechopen.com

Numbers displayed above are based on latest data collected.
For more information visit www.intechopen.com

Meet the editor

Associate professor Dr. Fathi Samir Soliman is an expert in petroleum refinery, especially production, treatment and recycling of lube oils and paraffin waxes with catalytic conversion and solvent extraction. He is also highly qualified in studying the morphology of Nano materials using high-resolution transmission electron microscopy and scanning electron microscopy. Since 2014, he has been a researcher at the physical refinery laboratory, Refinery Department, Egyptian Petroleum Research Institute (EPRI) where he is also a supervisor of the electron microscope laboratories in the Nano Technology Centre. Dr. Soliman is a lecturer at the Higher Institute of Applied Arts, and a member of the organizing committee of the Egyptian annual international conference on Petroleum, Mineral Resources and Development.

Contents

Preface

The book in your hands, *Paraffin – an Overview*, can teach you the basics of paraffin. It covers all the paraffin pathways from exploration to the applications. Taken as a whole, these five chapters aim to help academic researchers and those connected with the petroleum industry (such as petroleum engineers) to easily understand the value of paraffins as a petroleum byproduct. The chapters cover the separation, transportation, and application technology of paraffins and these chapters are presented in a careful and clear manner to provide an easy and enjoyable read. Data and discussions are presented by highly qualified experts and the text is edited to simplify the information. This book is valuable for those who are interested in making a connection between natural resources, science, and industry.

Fathi Samir Soliman
Egyptian Petroleum Research Institute,
Egypt

Chapter 1

Introductory Chapter: Petroleum Paraffins

Fathi Samir Soliman

1. Introductory

Waxes separated from petroleum are defined as the waxes present naturally in various fractions of crude petroleum [1]. Petroleum waxes are complex mixtures of hydrocarbons, amongst which are n-paraffin, branched chain paraffins and cyclo-paraffins in the range of C_{18}–C_{70} [2–4]. The quality and quantity of waxes manu-factured from crude oils depend on the crude source and the degree of refining to which it has been subjected prior to wax separation [5, 6].

Paraffin waxes constitute the major bulk of such waxes, the other two types; produced in comparative quantities; also command a good market because of their certain specific end uses [2]. The paraffin waxes are solid hydrocarbons at room temperature.

Slack wax is a refinery term for the crude paraffin wax separated from the solvent dewaxing of base stocks. Slack wax contains varying amounts of oil (rang-ing from 20 to 50 wt.%) and must be removed to produce hard or finished waxes [1, 7]. If the slack wax separated from residual oil fractions, the oil-bearing slack is frequently called petrolatum [8].

Petrolatum is a general name applied to a slightly oiled crude microcrystalline wax. It is semi-solid, jelly-like materials. Petrolatum is obtained from a certain type of heavy petroleum distillates or residues.

Ozokerite wax is naturally occurring mineral wax. It is also a microcrystalline wax. Ceresin is a microcrystalline wax; it is the name formerly given to the hard white wax obtained from fully refined ozokerite. Petroleum ceresin is a similar microcrystalline wax but separated from petroleum. Ceresin and petroleum ceresins appear to have the same composition, structure, physical and chemical properties [1].

1.1 Composition of petroleum waxes

Petroleum waxes are substance, which is solid at normal temperatures. Paraffin and microcrystalline waxes in their pure form consist only solid saturated hydrocar-bons. Petrolatum, in contrast to the other two waxes, contains both solid and liquid hydrocarbons. Petrolatum is semi-solid at normal temperatures and is quite soft as compared to the other two waxes.

Paraffin wax is a solid and crystalline mixture of hydrocarbons; it is usually obtained in the form of large crystals. It consists generally of normal paraffin rang-ing from C_{16} to C_{30} and may be higher. Proportions of slightly branched chain paraf-fin ranging from C_{18} to C_{36} and naphthenes; especially alkyl-substituted derivatives of cyclopentane and cyclohexane; are also present [1, 5, 8–10].

The average molecular weight of these paraffin waxes is about 360–420 [9, 11]. A paraffin wax melting at 53.5°C showed a space lattice having C—C bond length of 1.52°A, a C—C—C bond angle of 110°A, a C—H bond length of 1.17°A and an H—C—H bond angle of 105°A [12].

Microcrystalline waxes are obtained from the vacuum residue. The source for the production of microcrystalline wax is petrolatum or bright stock [13].

Microcrystalline waxes consist of highly branched chain paraffin; in contrast to the macrocrystalline; cycloparaffins and small amounts of n-paraffins and alkylated aromatics [1, 5, 9]. The actual chain length of the n-alkanes is approximately C_{34}–C_{50}. Long-chain, branched iso-alkanes predominantly contain chain lengths up to C_{70} [13].

The branched-chain structures of the composition C_nH_{2n+2} are found. Branched mono-methyl alkane, 2-methyl alkanes being found. As the position of the methyl group moves farther from the end of the chain, the amount of the corresponding alkane becomes smaller. The branched chains in the microcrystalline waxes are presented at random along the carbon chain, meanwhile in paraffin wax, they are located at the end of the chain [14].

The cyclo-alkanes, however, consist mainly of monocyclic systems. Monocyclopentyl, monocyclohexyl, dicyclohexyl paraffin and polycyclo paraffin are also found. Some microcrystalline waxes are mainly composed of multiple-branched isoparaffins and monocycloparaffins [1]. Moreover, non-hydrogenated micro waxes also mainly contain mono-cyclic and heterocyclic aromatic compounds [13].

Microcrystalline waxes have higher molecular weights (600–800), densities and refractive indices than paraffin waxes [1, 5, 6, 9].

1.2 Properties of petroleum waxes

1.2.1 Physical properties

Almost all, the physical properties of petroleum waxes are affected by the length of hydrocarbon chain, distribution of their individual components and degree of branching [10, 15].

Paraffin waxes are composed of 40–90 wt.% normal paraffins of about 22–30 carbon atoms and possibly higher, accordingly, they differ very little in physical and chemical properties. The remainder is C_{18}–C_{36} isoalkanes and cycloalkanes [5, 16]. Straight chain alkanes in the range from 20 up to 36 carbon atoms show transition points in the solid phase. Thus two modifications, stable at different temperatures and different crystal habits, are known [1].

Microcrystalline waxes contain substantial proportions of highly branched or cyclic hydrocarbons in the range from 30 to 75 carbon atoms [5, 6, 17].

Paraffin waxes, relatively simple mixtures, usually have a narrow melting range and are generally lower in melting point than microcrystalline waxes. They usually melt between 46 and 68°C. The melting point of paraffin waxes increases in parallel with molecular weight. The branching of the carbon chain, at identical molecular weights, results in a decrease in the melting point. Paraffin waxes can be classified according to the melting point to soft (lower m.p.) and hard (higher m.p.) paraffin waxes.

Microcrystalline waxes are more complicated so it melts over a much wider temperature range. They usually melt between 60–93°C and 38–60°C, respectively [6, 9, 10].

Oil content is a fingerprint of the quality of the wax. The method of determination depends upon the differential solubility of oil and wax in a given solvent.

Paraffin wax, microcrystalline wax and petrolatum have a different degree of affinity for oil content. Paraffin wax has little affinity for oil content. It may be taken as a degree of refinement. Fully refined wax usually has an oil content of <0.5%.

Microcrystalline waxes have a higher affinity for oil than paraffin waxes because of their smaller crystal structure. The oil content of microcrystalline wax is 1–4 wt.%, depending on the grade of wax [9].

1.2.2 Mechanical properties

The hardness and crystallization behavior of macrocrystalline paraffin waxes are interfered distinctly by their distribution width, average chain length and n-alkane content [18].

Hardness is the resistance against the penetration of a body (needle, cone or plunger rod) under a defined load, this body is made of a harder material than the substance being tested. To measure the hardness of paraffin waxes, penetration tests are widely accepted. It is a common feature of strength and hardness tests that the test specimens are subjected to short-time stresses [10].

The penetration test is the most widespread technique for determining the hardness and the thermal sensitivity of petroleum waxes. Macrocrystalline waxes change to a greater extent with temperature than that of microcrystalline waxes. An increase in oil content results in an increase in penetration values of both macro- and microcrystalline waxes [1].

1.2.3 Food grade properties

These properties concern waxes and petrolatums for food grade. Their potential toxicity could be attributed to aromatic residues. The latter are characterized directly by using UV spectra in the spectral zone corresponding to aromatics.

Each country has adopted its own code governing waxes, which come in contact with food and non-food grade [19, 20].

1.3 Crystal structure of petroleum waxes

The class of organic crystals represents a broad range of geometries, including needles, plates, cubes, rods, prisms, pentagons, octagons, hexagons, rhomboids and pyramids. Each of these forms results from crystallization from a solution. The geometry of the crystals formed is determined by the solute/solvent interaction and the physical conditions of the system (e.g., temperature, pressure and mechanical mixing).

One interesting characteristic of crystals is that they can form a variety of shapes, which are due to the environmental conditions under which they form. They can be large or small, extend long distances or short, be well-defined or diffuse; in short, they can display an impressive array of forms. It is this variety of form upon which crystal modifiers are intended to take advantage [21].

All petroleum waxes are crystalline in some degree and it is possible to classify waxes in terms of the type of crystals formed, when the wax crystallizes out of solution.

1.3.1 Macrocrystalline waxes (paraffin waxes)

The paraffin crystals appear in three different forms: plates, needles and mal shapes; the latter are small size, undeveloped crystals, which often agglomerates. The conditions for the formation of these shapes have been studied by many researchers. They have come to the following conclusions:

1. The three crystal forms of paraffin waxes depend on both the conditions of the crystallization process and the chemical composition of the wax.

2. Plate crystals are obtained from lower boiling points paraffinic distillates, while the needle and mal-shaped crystals are obtained from the higher boiling points ones and from vacuum residues.

3. For a given molecular weight limit, the higher melting point constituents crystallize in plate type in which the crystals are hexagonal plate. The low-melting ones crystallize in needles while the medium-melting ones crystallize in mal shapes.

4. Normal paraffin crystallize in plates. Needle crystals contain both aliphatic and cyclic hydrocarbons, while mal-shaped crystals are characterized by their content of branched hydrocarbons.

5. Low-cooling rates during crystallization will result in large crystals for both plate and needle forms, while the crystal growth for mal-shaped crystals is very slight.

6. The solubility of paraffin in a solvent is inversely proportional to their melting points. In the presence of solvent, wax mixtures begin to crystallize at relatively low temperatures in the form of plates followed by mal-shaped crystals. However, the constituents crystallizing in needles are more soluble than those crystallizing in plates. Therefore, needles crystals will appear only at lower temperature and higher concentrations

7. Plate crystals can readily be transformed into needle and mal-shaped crystals. Under appropriate conditions, the needle crystals can be transformed into mal-shaped crystals [10].

Normal paraffin, C_{17}–C_{34}, may exist in three and possibly four crystal forms. Near the melting point, hexagonal crystals are the stable form. At somewhat lower temperatures, the odd-numbered from C_{19} to C_{29} are orthorhombic, even numbered ones from C_{18} to C_{26} is triclinic and those C_{28}–C_{36} is monoclinic [22, 23].

1.3.2 Microcrystalline waxes

Both n-paraffin and isoparaffins crystallize in needle forms; they differ in that the latter does so at all temperatures, while higher temperatures are required for the former. The needle form of the isoparaffins differs from that of ceresins or paraffin waxes containing ceresins, in that the crystals of former are large and loose, while those of the latter are extremely small and dense [14].

Microcrystalline waxes may contain substantial percentages up to 30% of paraffin which, when separated, crystallize well as high-melting macro crystalline or paraffin wax. The microcrystalline wax material interferes and imposes its crystallizing habit on the other material [16, 24].

Although the classification of petroleum waxes into macro crystalline and microcrystalline waxes on the basis of crystal size is valid to a great extent, there is no sharp line separating the two groups. Indeed, there is a large group of waxes that could fall in either classes and these waxes are called intermediate waxes, blended waxes, mal-crystalline waxes and semi-microcrystalline waxes. But semi-microcrystalline wax adopted [17].

1.4 Manufacture of petroleum waxes

The manufacture of petroleum waxes is closely related to the manufacture of lubricating oils. The raw paraffin distillates and residual oils contain wax and they are normally solid at ambient temperature. Removal of wax from these fractions is necessary to permit the manufacture of lubricating oil with a satisfactory low pour point. Manufacture of petroleum waxes includes the following technological processes:

- Production of slack waxes and petrolatums by dewaxing petroleum products.

- Refining of the wax products.

- Deoiling and fractional crystallization.

- Percolation process.

- Hydrofinishing process.

- Acid treatment.

- Adsorption process.

1.5 Applications of petroleum waxes

As the consumption of wax products in the world wax market increases; especially for food, pharmaceutical and cosmetic grades and specialty wax; the increase of profitability of wax production will lie on the improvement of blending and modification techniques for macro- and microcrystalline waxes as base materials as well as the development and applications of new wax products [25].

Petroleum waxes are based in a wide variety of applications. Some of its most important applications were used in industry such as, paper industry, household chemicals, cosmetics industry, dental industry, match industry, rubber industry, building constructions, electrical industry, inks industry and powder injection molding industry beside that of hydrogen production and energy storage applications [4, 10, 26–30].

2. Conclusion

Fractions of petroleum wax can be achieved to separate more than one type of paraffin wax such as macrocrystalline and microcrystalline waxes, the waxes characterization such as carbon number, hardness, crystal shape, composition and molecular weight depend on the condition of separating the wax, paraffin wax act like a joker in different industries such as inks, papers, cosmetics and ceramic fabricating using powder injection molding industry.

Author details

Fathi Samir Soliman
Egyptian Petroleum Research Institute, Cairo, Egypt

*Address all correspondence to: fathisamir@gmail.com

IntechOpen

References

[1] Mazee W. Modern Petroleum Technology. Great Britain: Applied Science Publishers Ltd., on behalf of The Institute of Petroleum; 1973. p. 782

[2] Prasad R. Petroleum Refining Technology. Delhi, India: Khanna; 2000

[3] Gupta A, Severin D. Characterization of petroleum waxes by high temperature gas chromatography-correlation with physical properties. Petroleum Science and Technology. 1997;15(9-10):943-957

[4] Bennett H. Industrial Waxes. New York: Chemical Pub Co; 1975

[5] Letcher C. Waxes. John Wiley & Sons New York. 1984. pp. 466-481

[6] Avilino S Jr. Lubricant Base Oil and Wax Processing. New York: Morcel Dekker, Inc.; 1994. pp. 17-36

[7] Guthrie VB. Petroleum Products Handbook. McGraw-Hill; 1960

[8] Concawe. Petroleum Waxes and Related Products. Boulevard du Souverain, Brussels, Belgium; 1999

[9] Gottshall R, McCue C, Allinson J. Criteria for Quality of Petroleum Products. London, Great Britian: Applied Science Publishers Ltd. On behalf; 1973

[10] Freund M. et al. Paraffin Products Properties, Technologies, Applications. 1982. p. 14

[11] Nakagawa H et al. Characterization of hydrocarbon waxes by gas-liquid chromatography with a high-resolution glass capillary column. Journal of Chromatography A. 1983;260:391-409

[12] Vainshtein B, Pinsker Z. Opredelenie polozheniya vodoroda v kristallicheskoi reshetke parafina. Doklady Akademii Nauk SSSR. 1950;72(1):53-56

[13] Meyer G. Thermal properties of micro-crystalline waxes in dependence on the degree of deoiling. SOFW journal. 2009;135(8):43-50

[14] Levy E et al. Rapid spectrophotometric determination of microgram amounts of lauroyl and benzoyl peroxide. Analytical Chemistry. 1961;33(6):696-698

[15] Kuszlik A et al. Solvent-free slack wax de-oiling—Physical limits. Chemical Engineering Research and Design. 2010;88(9):1279-1283

[16] Corson B. In: Brooks BT, Kurtz SS Jr, Boord CE, Schmerling L, editors. The Chemistry of Petroleum Hydrocarbons. Vol. Ill. 1955. pp. 310-312

[17] Ferris S. Petroleum Waxes: Characterization, Performance, and Additives. New York, USA: Technical Association of the Pulp and Paper Industry; 1963. pp. 1-19

[18] Meyer G. Interactions between chain length distributions, crystallization behaviour and needle penetration of paraffin waxes. Erdöl, Erdgas, Kohle. 2006;122(1):16-18

[19] Hopkins TD, N.C.F.P. Analysis, the costs of federal regulation. National Chamber Foundation. 1992

[20] USP, U.P. 34, NF 29. The United States pharmacopeia and the National formulary. Rockwille, MD: The United States Pharmacopeial Convention; 2011

[21] Becker J. Crude Oil Waxes, Emulsions and Asphaltenes. Tulsa, OK, USA: Penn Well Publishing Company; 1997

[22] Smith A. The crystal structure of the normal paraffin hydrocarbons.

The Journal of Chemical Physics. 1953;**21**(12):2229-2231

[23] Ohlberg SM. The stable crystal structures of pure n-Paraffins Contalmng an even number of carbon atoms in the range C_{30} to C_{36}. The Journal of Physical Chemistry. 1959;**63**(2):248-250

[24] Higgs P. The utilization of paraffin wax and petroleum ceresin. Journal of the Institution of Petroleum Technology. 1935;**21**:1-14

[25] Zaky MT et al. Raising the efficiency of petrolatum deoiling process by using non-polar modifier concentrates separated from paraffin wastes to produce different petroleum products. RSC Advances. 2015;**5**(88):71932-71941

[26] Maillefer S, Rehmann A, Zenhaeusern B. Hair wax products with a liquid or creamy consistency. Google Patents. 2011

[27] Saleh A, Ahmed M, Zaky M. Manufacture of high softening waxy asphalt for use in road paving. Petroleum Science and Technology. 2008;**26**(2):125-135

[28] Zaky M, Soliman F, Farag A. Influence of paraffin wax characteristics on the formulation of wax-based binders and their debinding from green molded parts using two comparative techniques. Journal of Materials Processing Technology. 2009;**209**(18-19):5981-5989

[29] El Naggar AM et al. New advances in hydrogen production via the catalytic decomposition of wax by-products using nanoparticles of SBA frame-worked MoO_3. Energy Conversion and Management. 2015;**106**:615-624

[30] Mohamed NH et al. Thermal conductivity enhancement of treated petroleum waxes, as phase change material, by α nano alumina: Energy storage. Renewable and Sustainable Energy Reviews. 2017;**70**:1052-1058

Chapter 2

Wax Chemical and Morphological Investigation of Brazilian Crude Oils

Erika C.A. Nunes Chrisman, Angela C.P. Duncke,
Márcia C.K. Oliveira and Márcio N. Souza

Abstract

The waxes in petroleum can precipitate and form unwanted gels and deposition when exposed to low temperatures. The idea of this chapter is to approach methods of quantification and physicochemical and morphological characterization of waxes and how this information can help in understanding this deposition. Information such as the quantity of waxes and the chemical structures in the oil is fundamental to predict the possible deposition and its ability to aggregate with other crystals. For example, the knowledge about the wax morphology may contribute to explain the nucleation and growth of the deposits. The polarized light microscopy, the most common technique to visualize wax crystals, and the bright-field microscopy, the most simple technique, able to show crystal details that have not been seen on the polarized light, was used.

Keywords: waxes, crude oil, quantification, characterization, microscopy, DSC

1. Introduction

Petroleum is a complex mixture of hydrocarbons of varying nature and small fractions of nitrogen, oxygen, sulfur, and metal compounds. At room temperature, petroleum can be gas, liquid, and/or solid, being considered as gases and solids dispersing in a liquid phase [1]. Under high temperature and pressure, as encountered at reservoirs (e.g., 8000–15,000 psi and 70–150°C), Newtonian rheological behavior prevails, whereas at low temperatures the pseudoplastic behavior is commonly found [2].

A large portion of the Brazilian oil production comes from offshore fields, from the pre-salt layer. These oils have high levels of waxes, which are alkanes (linear or branched) encompassing carbonic chains of 15–75 carbons [3, 4]. This class of compounds has a high precipitation potential, due to the low sea temperatures (about 4–5°C) [5–7]. In temperatures below the wax appearance temperature (WAT), the wax crystallization takes place with subsequent deposition [2]. The wax deposition is dominated by the molecular diffusion mechanism [8] in which the waxes initially precipitate at the cold pipeline walls and subsequently generate a radial gradient of precipitation causing deposit [9, 10]. This can lead to a strong waxy crystal interlocking network, which causes pipeline clogs and dramatically affects the rheological fluid behavior [9, 11–13].

IntechOpen

Gelation and deposition problems, leading to increases in yield stress and losses in production, are probably connected to wax morphology. This chapter aims to show some techniques to characterize the structure and morphology of wax crystals based on four pre-salt Brazilian crude oils, all provided by Petrobras, under different shear conditions, aging times, and temperatures. In addition, some physico-chemical characterization techniques are discussed as density, viscosity, and SAP (saturated, aromatic, and polar). The wax quantification is the harder part of the study of crude oils, due to the petroleum complex matrix, which can cause complications related to the wax crude oil separation; however, through differential scanning calorimeter (DSC) measurements, it is possible to obtain a precipitated wax content as well as through some American Society for Testing and Materials (ASTM), Universal Oil Products Collection (UOP), gas chromatography (GC), and others.

2. Morphological characterizations

Due to the petroleum multicomponent nature, the wax precipitation occurs heterogeneously, and resins and asphaltene molecules, inorganic solids, and corrosion products, among others, can behave as nuclei for the phenomenon, enhancing the flow assurance issue [14].

Waxes crystallize into basically orthorhombic and hexagonal shapes. The orthorhombic form is needle-shaped, and it is found in crudes with high waxy content [15, 16]. Crystallization kinetics and crystal morphology can be highly affected by some recognized factors, such as cooling rate [13, 17–23], carbonic chain nature (branched or linear and average length) [21], resins and asphaltene content [2, 7, 24, 25], and shear rate [16, 26–28].

The polarized light (PL) optical microscopy is the fundamental technique for wax crystal examination [24]. According to [29] it allows verifying the anisotropic optical behavior of crystalline materials, named birefringence. This technique uses two cross polarizers. When the light beam passes through crystalline structures, as wax crystals, the polarized light plane is altered generating a visible image pattern. On the other hand, isotropic structures, which do not exhibit the same level of organization, are not able to modify the light plane. Apart from PL microscopy, the bright-field (BF) microscopy regards another important technique for wax crystal visualization. The procedure is very simple, and no artifacts are employed in the optical path.

Figure 1 shows BF and PL micrographs of P1 Brazilian crude oil, for the same point of the coverslip, at 25°C, as received, i.e., without any thermal treatment. All

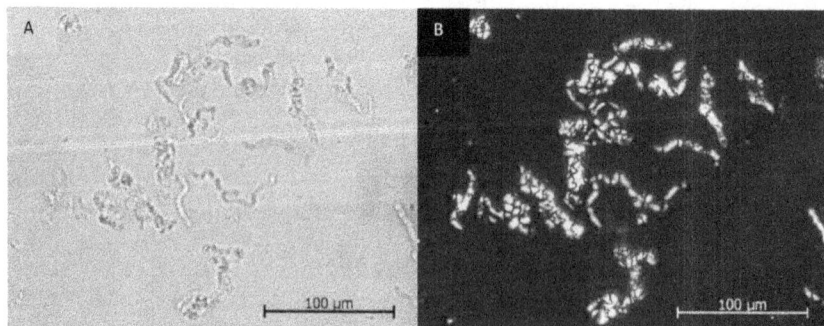

Figure 1.
(A) BF and (B) PL micrograph of P1, for the same point of cover slip at 25°C, as received.

the aliquots of crude oil in this chapter were observed on optical microscope Axio Vert 40 MAT (Carl Zeiss).

The BF technique (**Figure 1a**) provides lower contrast than PL technique (**Figure 1b**); however, it can be seen that in BF micrographs the wax crystal is continuous, i.e., the structure appears and integrates, without rupture. On the other hand, PL micrographs show "dark cracks," i.e., the wax crystals do not appear entirely. These "dark cracks" can be attributed to two factors: first, amorphous or low crystallinity regions due to the presence of impurities and second, due to light extinction positions, related to the parallel orientation of polarizers and the crystal organization, i.e., no light is deflected by the sample [30]. Therefore, much attention should be taken to make length measurements in crystals observed by PL technique. According to these results, to determine the size and crystal shape (as verified by BF) can be critical to avoid erroneous measurements. In this work, the length measurements were performed on images obtained by BF, but the PL images are shown due to easy observation.

Another characteristic of wax crystals that can be seen in **Figure 1a** is a roughened surface. The roughness, as well as the tortuosity of wax crystals, can be attributed to a heterogeneous nucleation and growth, by the presence of asphaltenes, resins, and different wax chain lengths or the presence of isocycle [24, 31].

In order to characterize the wax morphology and crystals length in dependence of temperature and shear, a continuous cooling protocol was performed (**Figure 2**). Initially, the thermal history removal of 100 mL of each oil was carried out by heating the samples for 2 h at 80°C in a circulating oven model 400-3ND (Ethik Technology). This condition is sufficient to dissolve all wax present in the crude oil and prevent that the wax crystal formation was not influenced by pre-existing nuclei [32, 33]. Secondly, the samples were transferred to a jacketed Becker coupled to a circulation bath (Haake Phoenix II-C25P - Thermo Scientific). Then, the cooling step was carried out quiescently or in presence of shear (mechanical agitation 250 rpm on RW20 Digital IKA) for 80–5°C. The cooling rate was 0.5°C/min. **Figure 2** shows the influence of shear on waxy crystal growth of P1–P4 paraffinic oil comparing the PL micrographs of tests carried out at 5°C, on quiescent and shear cooling conditions.

It was verified that experiments performed with quiescent condition (**Figure 2A–D**) were characterized by large crystals and cluster of crystals when compared with experiments carried out with shear condition (**Figure 2E–H**). The

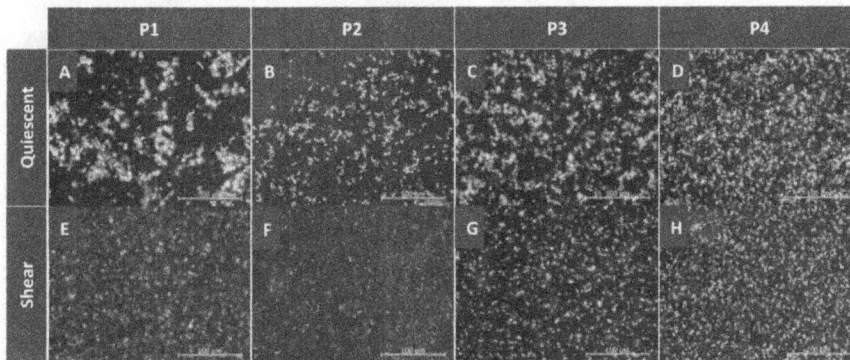

Figure 2.
PL micrographs of test performed at 5°C on quiescent (A–D) and shear (E–H) conditions of waxy crude oils P1–P4.

researchers carried out by [2, 16, 34] show that under quiescent conditions, the waxy crystals were characterized by extended and continuous particles. The formation of extended and continuous particles allowed a colloidal network that embodies the oil itself. Probably, the gel would have a high shear modulus, in order to the side-by-side interactions between particles. Under the shear condition, the lateral growth of the individual crystals is constricted. However, extended particles are not observed, and consequently, these particles lost their interconnectivity.

The wax crystals presented in waxy crude oils (**Figure 2**) are elongated. According to [16], waxes precipitated in crude oil tend to crystallize in an ortho-rhombic structure, which is characterized by an elongated structure. Evidently, the crystals of **Figure 2** (and in detail in **Figure 1**) are not linear (needle-like). The sinuosity and tortuosity are probably due to the presence of impurities during nucleation and crystal growth processes [2, 21]. [2] analyzed the aspect ratio, which is the ratio between the length and the width of a crystal. Based on aspect ratio value, it is possible to verify how the structure is elongated. The values of average aspect ratio, at 5°C, of samples P1, P2, and P3, are 5.5, 6.2, and 5.0, respectively, legitimizing the elongated characteristic. P4 sample has a 4.0 aspect ratio value, which indicates that the crystals are less elongated than other samples.

Table 1 shows the average length and width of crystals to waxy crude oils P1–P4 in function of temperature for 30, 10, and 5°C, for quiescent and shear conditions, and shows the average percentage of crystal growth between both cooling conditions.

For quiescent conditions, it is possible to note the crystal length increases between 10 and 5°C; however, for shear conditions, the length becomes basically stationary at these temperatures. This behavior could be attributed to a possible crystal breakage by the shear, which prevents the crystals from becoming large. The average percentage of growth between quiescent and shear conditions increases with the temperature decrease. For 30°C the crystals obtained in quiescent cooling are about 12.4% higher than that obtained by shear conditions. At 5°C this differ-ence reaches 25.1%. On the other hand, the crystal width underwent an effective action of the shear, being about 22.3% less wide than those obtained in quiescent conditions.

To illustrate the **Table 1**, **Figure 3** shows PL micrographs of P3 obtained at 30, 10, and 5°C during quiescent cooling. This condition resembles the operational shutdowns when crude oil is cooled. As expected and discussed above, the concen-tration and size of wax crystals increase with the decrease in temperature. Since the solubility of high molecular weight waxes decreases sharply with the decrease in temperature, they precipitate out and crystallize. This result indicates that in low temperatures, it is more probably to have problems of flow assurance due to pipe-line blockage occasioned by wax crystal depositions and to the formation of a high-strength gel, characterized by yield stress [35–37].

	T (°C)	P1		P2		P3		P4		Average Growth (%)
		Quiescent	Shear	Quiescent	Shear	Quiescent	Shear	Quiescent	Shear	
Length (μm)	30	12.0	8.3	6.2	5.3	10.2	10.0	8.2	8.0	12.4
	10	14.6	12.7	14.6	9.0	13.3	11.3	9.6	8.9	18.4
	5	16.2	12.8	14.9	8.9	15.4	11.3	10.4	9.1	25.1
Width (μm)	30	2.3	1.5	2.4	1.5	2.3	1.9	2.2	2.0	24.7
	10	3.0	2.5	2.7	1.8	3.2	2.6	2.6	2.3	20.0
	5	3.1	2.4	2.9	1.9	3.3	2.6	2.7	2.4	22.3

Table 1.
Length and width of crystal's average and growth percentage.

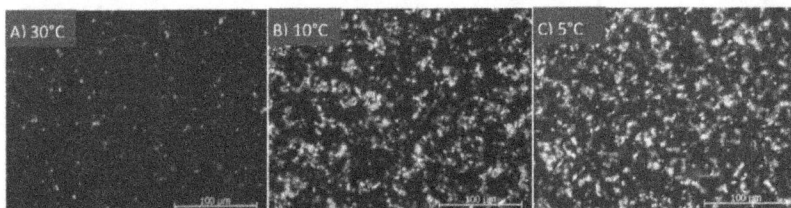

Figure 3.
PL micrographs of P3 obtained at (A) 30°C, (B) 10°C, (C) 5°C and during quiescent cooling.

Another common factor studied on precipitation and morphology of waxy crystals is the aging time, which represents the influence of the time at a certainly constant temperature on the crystal wax. PL micrographs in **Figure 4** show the influence of 1 h aging time at temperatures 40, 20, and 5°C, for P4. To study the aging time influence, first, the thermal history was removed. The samples were transferred to the jacketed Becker coupled to a circulation bath at 80°C and then started the cooling steps (80–40°C; 80–20°C or 80–5°C). When the temperature arrives 40, 20, or 5°C, the samples were kept cool for 1 h at this temperature. The cooling rate was 0.5°C/min.

It was verified that the aging time favored the increase of crystal length and appearance of large clusters. This result can be attributed to the Ostwald ripening of wax crystals, a mechanism by which the large crystals grew at the expenses of smaller crystals of higher energy. Furthermore, oil uptake can also change the wax crystal distribution, leading to larger and softer wax crystals that can interpenetrate increasing intermolecular interactions between crystals [11, 37, 38].

Table 2 shows the wax crystal's average length at t = 0 h and after 1 h (t = 1 h) at temperatures 40, 20, and 5°C, as well as the crystal growth percentage in function of aging time.

Analyzing **Table 2**, at 40°C the oils P1 and P3 show an increase of about 26.3% in the length of the crystal after 1 h in an isothermal condition. Under these same conditions, P2 shows a growth of almost 80.0%. P2 has the WAT at 42.1°C (see 4. Wax quantification), and consequently, there is no visible crystal on microscope when the temperature just arrives at 40°C. For this reason, the crystal size, in this

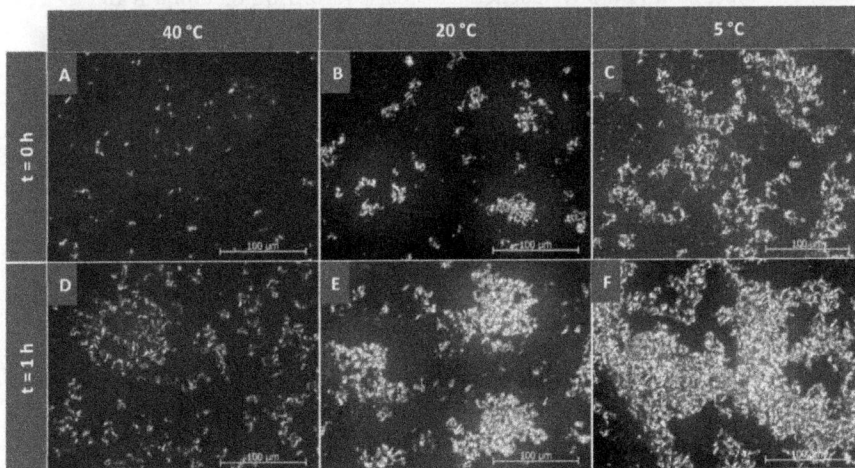

Figure 4.
PL micrographs of tests carried out with P4, at t = 0 h for 40°C (A), 20°C (B) and 5°C (C); and after 1 h at 40°C (D), 20°C (E) and 5°C (F).

Oil	40 °C			20 °C			5 °C		
	t = 0 h (μm)	t = 1 h (μm)	Growth %	t = 0 h (μm)	t = 1 h (μm)	Growth %	t = 0 h (μm)	t = 1 h (μm)	Growth %
P1	8.2	11.2	26.7	13.0	15.9	17.8	16.1	18.3	12.4
P2	*	4.8	79.0	7.3	10.2	28.1	15.5	17.2	9.6
P3	7.9	10.6	26.0	11.5	14.3	19.6	15.1	16.1	6.1
P4	4.9	8.9	45.5	9.3	12.6	26.0	10.7	12.3	13.2

No crystals were present. The crystal size of 1.0 μm was convened because it is the detection limit of the microscope.

Table 2.
Average wax crystal length at t = 0 h and t = 1 h at 40, 20, and 5°C and crystal's growth percentage.

case, was considered 1.0 μm, the microscope detection limit. However, after 1 h at 40°C, this sample presents small crystals of about 4.8 μm. Evaluating a percentage of growth at 20 and 5°C, a reduction is noticed. The wax crystals seem to grow more significantly at elevated temperatures. In t = 0 at 5°C, the wax crystal has a large size due to the temperature decrease, and after 1 h in an isothermal condition, the wax crystal grows little, i.e., its sizes do not "double" as at 40°C. A smaller variation was noted between the sample growth percentages at 5°C. This temperature is close to that observed in the production fields. After 1 h at 5°C, the wax crystals are 10.3 ± 2.8% higher than when the temperature just arrives 5°C. Generalizing this information and transferring it to offshore production fields, after a 1-h stop with the oil at 5°C, the crystals can grow about 10%. Of course, this is a hypothetical condition because it is impossible to happen, since the cooling rate in the fields is smaller than that used in this study, which can result in greater wax crystals.

3. Physicochemical characterization

Due to the complex matrix that is the petroleum itself, the physicochemical characterization is very relevant in order to address a proper comparison between the microscopic images, which is a very useful tool in the wax crystal morphology study. The most common physicochemical characterization techniques are:

- Density: measured mainly by ASTM-D7042. By density (at 60°F = 15.6°C) it is possible to obtain the °API following Eq. (1). °API is the most general classification at petroleum industry:

$$°API = \frac{141.5}{\rho} - 131.5 \tag{1}$$

- Viscosity: can be also determined by ASTM-D7042 on a viscometer or by rheological tests.

- Saturated, aromatic, resin, and asphaltene (SARA): can be determined mainly by Clay-Gel, according to ASTM D2007, thin layer chromatography with flame ionization detection (TLC-FID) according to IP-469, or by high-performance liquid chromatography (HPLC) according to IP-368. In this work, SARA content was obtained by TLC-FID using the IATROSCAN MK-6 (NTS International), for all paraffinic crude oils.

- SAP: this characterization is less specific than SARA because resins and asphaltenes are considered together as polars. The SAP contents were determined by a liquid chromatography separation composed by silica gel column 60 (2.5 g silica, 0.063–0.200 mm) from Merck, which was used to determine the SAP content. The column was heated for 10 hrs at 120°C for activation. Fractions were eluted with 10 mL n-hexane for saturated, 10 mL of n-hexane/dichloromethane (8:2) for aromatic, and dichloromethane/methanol (9:1) for polar fractions. Residual solvents were submitted to a rotary evaporation. This technique was employed only for non-paraffinic (NP).

- WAT: this is one of the main characterizations when working with waxy crudes, because it gives an idea of the precipitation potential of the oil and ideas about the wax type. A wide range of techniques can be used to determine WAT, as microscopy, rheology, and near-infrared spectroscopy (NIR), but the most used is DSC. In this work, measurements were performed using Nano DSC differential scanning calorimeter (TA Instruments). The samples were heated from room temperature to 80°C, at 2°C/min. Then they were held for 15 min at 80°C, following by a cooling step from 80–4°C, at 0.5°C/min. Kerosene was used as the reference. Before measurements, samples were homogenized and kept under vacuum for degasification for at least 30 min. A volume of 300.0 µL of crude was used.

- Gas chromatography: this technique is employed to characterize the carbon number distribution of petroleum waxes and the normal and non-normal hydrocarbons. It is oriented by ASTM D5442-17. In this work the GC evaluated the carbon distribution up to C36.

   Table 3 presents some physicochemical characterization of the four paraffinic P1–P4 and NP oils used as reference of wax absence, also provided by Petrobras. All crude oils have relatively similar values of density. The paraffinic samples are considered medium oils, while NP is classified as heavy oil according to the °API scale. The viscosity varies greatly between samples, with P1 and P3 being the less viscous. P4 exhibits the highest viscosity at 20°C, being 100 times greater than the lower one (P3). Non-paraffinic petroleum classified as heavy oil also has high viscosity (896.8 mPa.s).

   WAT is defined as the onset temperature, that is, the intersection point of the baseline and the tangent line of the inflection point of the exothermic peak [4, 39, 40]. In crude oils, it is common to observe two exothermic events (peaks). WAT depends on the concentration and molecular weight of waxes and the chemical characterization of hydrocarbon matrix [41]. Due to the oil complexity, the

Oil	ρ (g/cm³)	°API	µ (mPa.s)	WAT (°C)		Oil Composition (wt. %)			
				1° event	2° event	Sat.	Aro.	Res.	Asp.
P1	0.88	28.2	153.6 ± 1.1	46.6	25.7	54.0	24.0	22.0	<0.50
P2	0.91	23.3	427. ± 1.0	42.1	23.5	53.1	25.6	21.1	<0.17
P3	0.89	27.2	113.3 ± 1.3	46.8	25.0	63.1	18.2	18.6	<0.50
P4	0.91	23.6	11279.3 ± 178.9	53.1	29.4	40.4	16.2	42.7	0.65
NP	0.93	19.4	896.8 ± 2.5	37.9	19.2	57.6	28.9	13.4*	
*SAP - polars									

Table 3.
Physicochemical characterizations.

values of the peaks are around 50°C for the first exothermic event and 25°C for the second; [16, 42] assign the first peak to a liquid-liquid transition and the second to liquid-solid transition. However, in this paper, the authors believe that each exothermic event refers to a different group of waxes according to the chain length. [43, 44] declare that n-alkanes with similar carbon numbers can co-crystallize with the longer n-alkane chains.

Figure 5 shows the thermal curves for all samples obtained by Nano DSC. All oils have at least two well-defined exothermic peaks. It is possible to note a great similarity between the WAT values and the intensity of the exothermic peaks in the curves of the oils P1 and P3. However, the saturated values are quite different (**Table 3**). P1 has the 54.0 wt% and P3 has the 63.1 wt%, the highest values between the samples. Nevertheless, we must keep in mind that not all saturated content refers to wax; thus, these differences between saturated content among the oils do not represent the real wax content.

Continuing the analysis of **Figure 5**, it is noted that P2 was characterized by the lower WAT values and P4 shows the higher (**Table 3**), which may be an evidence that the P2 is composed by short waxy chains and P4 has the longest. According to [36] the larger the carbon chain size, the higher the crystallization temperature. Moreover, the first peak of P2 is barely evident which can be a sign of less wax content. P4 has a second peak very evident, that is, this oil may contain the higher wax content. However, P4 has the smallest crystals, as discussed before, being on average 35% smaller than the others are. According to the P4 higher WAT value, large crystals were expected. Senra et al. [45] suggest a co-crystallization between chains with different carbon numbers and with other compounds, affecting the crystal morphology. According to [46] the co-crystallization weakens the crystal structure and disfavors large crystal formation. This is a plausible hypothesis, since according to SARA, P4 has 42.7 wt% of resins and the higher content of asphaltenes (0.65 wt%).

Another curious fact is a possible third peak at temperatures just below the second, especially for P2 and P4. This peak may represent a third population of waxes, and as far as we know, it was never reported in conventional DSC analyses. Possibly this third peak is related to a group of very-short-chain waxes. Based on this observation, it is verified that the Nano DSC technique presents greater sensitivity to enthalpy variations. In the conventional DSC technique, this third peak may be masked with the second. According to [19] the conventional DSC is not

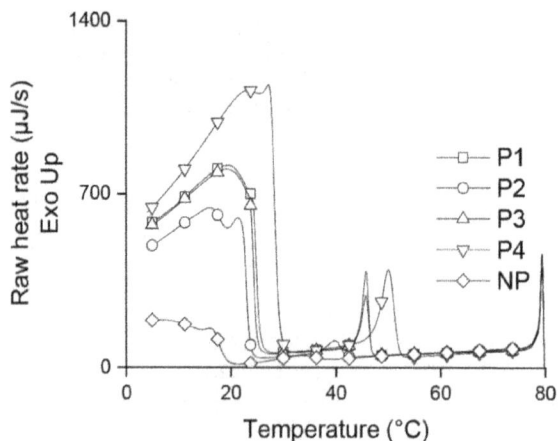

Figure 5.
P1-P4 and NP thermal curves behavior.

sufficiently sensitive to identify WAT for samples with low wax contents; however, the Nano DSC shows two slight baseline variations for NP sample, even in a low cooling rate (0.5°C/min). These peaks are very low if compared to other oils due to the non-paraffinic characteristic of NP, but their presence confirms the sensitivity of the equipment.

Figure 6 shows the GC graphs of the crude oils P1–P4 and their respective extracted waxes through the UOP46–85 method (see Section 4). It is possible to note that the values obtained for the GC of the crude oil (white bars) are dispersed

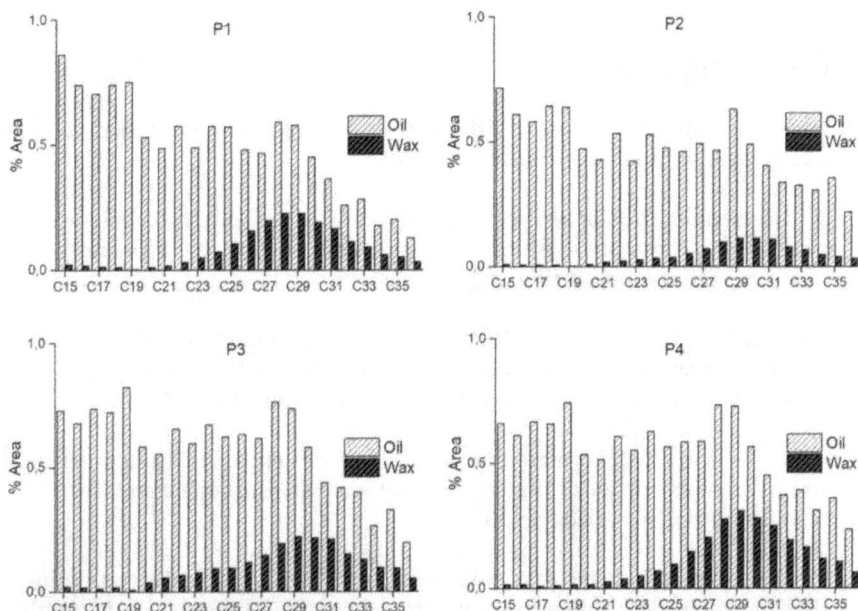

Figure 6.
Carbon number distribution for P1-P4 crude oil.

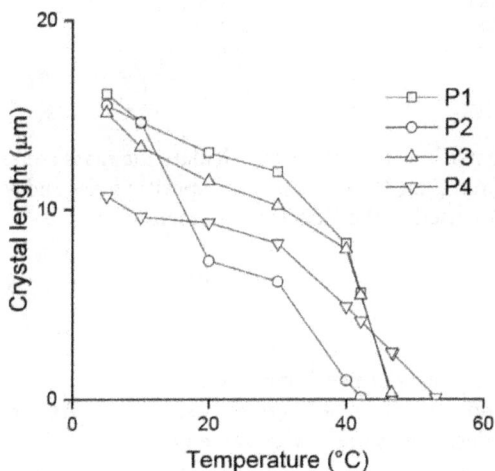

Figure 7.
P1-P4 crystal length versus temperature.

and have a tendency of decrease after around C30. This behavior can be attributed to the complex matrix of the oil itself. However, the carbon distribution number obtained from the extracted wax fraction from each oil (dark bars) has a more plausible chain distribution. For all oils, there is a chain predominance around C30.

Figure 7 shows the crystal length versus temperature for P1–P4. The first experimental point of the curves is the respective WAT values. This graph is presented in order to analyze the growth tendency of the wax crystals as a function of the temperature reduction, as a way to summarize the information previously discussed.

4. Wax quantification

The wax quantification is more difficult to develop than the other characterizations. However, some techniques are available:

- GC: as mentioned on 3. Physicochemical characterization, this technique is employed to characterize the carbon number distribution. In this work the GC evaluated the carbon distribution up to C36.

- Nuclear magnetic resonance (NMR) correlation: presented by [47], estimates the wax content of crude oil and their fractions by H NMR spectroscopy. The method shows good fit for oils with boiling range from 340 to 550°C.

- UOP 46–85 method: estimates the wax content of the crude oil and is defined as the mass percentage of precipitated material when an asphaltene-free sample solution is cooled to −30°C.

- DSC integration baseline: is possible to obtain the total thermal effect of the wax precipitation (Q) by integrating the area of the exothermic peaks. With this value, the wax precipitated concentration (c_w) can be determined following Eq. 2 [48]. $\overline{Q}$ is defined as 210 J/g, a constant thermal value of wax precipitation for crude oils with an unknown molecular structure [49]. WAT is the WAT temperature itself, and T_f is the final temperature, in this work, 4°C:

$$c_w = \frac{\int_{T_f}^{WAT} dQ}{\overline{Q}} = \frac{Q}{\overline{Q}} \tag{2}$$

By means of simple math, it is possible to calculate the mass content of precipitated waxes (w) as shown in Eq. (3), where ρ is the specific mass and V_e is the experimental volume used to the DSC measurement:

$$w = \frac{c_w}{\rho.V_e}.100 \tag{3}$$

The percentages by mass of precipitated wax obtained by the DSC integration baseline show 3.1 and 2.9 wt% for P1 and P3, respectively. As cited before these oils have many similarities. P2 has the lowest value (2.2 wt%) and P4 has 4.7 wt% of precipitated waxes. However, by the UOP 46–85 method, the wax contents in mass percentage obtained were 3.7 ± 0.3 for P1, 5.7 ± 0.4 for P2, 5.0 ± 0.1 for P3, and 3.6 ± 0.2 for P4. In general, these values are at the same range of the values

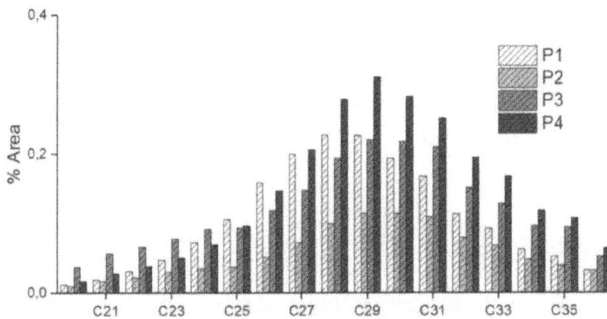

Figure 8.
Carbon number distribution for P1-P4 crude oil.

obtained by DSC integration baseline, but they are not in agreement with the values obtained by this same technique. The UOP 46–85 method is a traditional way of wax estimation by very steps extractions, as well as time-consuming, lots of chemicals and solvents. These many delicate steps have great chances to produce erroneous results if not done properly [47].

Figure 8 shows the carbon number distribution, obtained through GC, only for the extracted waxes by means of UOP method. As determined by DSC integration baseline, P2 has the lowest percentage of waxes, and P4 has the highest. This can be observed again on the GC graph. According to [50] the GC and DSC analyses can be used to quantify wax content of crude oils showing reasonable agreement, but wax precipitation technique, as UOP method, must be corrected due to the presence of trapped crude oil in the precipitated solid wax crystal.

5. Conclusions

The polarized light microscope is the most used technique to visualize wax crystals; however, bright-field microscopy shows crystal details that are not seen on the polarized light. The wax crystals observed have elongated structure, but they are not linear, i.e., not needle-shaped. They have superficial roughness attributed to the presence of crystallization interferers such as asphaltenes, resins, organic solids, and different carbon chain sizes. The gradual temperature decrease favors the length crystal increases, as well as the increase in the quantity and size of the agglomerates. Under shear conditions, crystals were observed around 25% smaller and in less quantity than under quiescent conditions. In addition, shearing promotes crystal breakage at very low temperatures. The aging time of the oil favors the crystal growth more drastically at higher temperatures (around 45% after 1 h at 40° C) than in low temperatures (around 10% after 1 h at 5°C), as well as the formation of agglomerates. P4 shows the higher content of precipitated waxes by means of DSC integration baseline and GC analysis, but their crystals were smaller, possibly due to the higher polar content. The DSC integration baseline is in accordance to the GC result to wax content determination; however, the UOP method is in disagreement. Another characteristic observed about Nano DSC was the great sensitivity to obtain WAT values. This technique can identify a possibly third peak precipitation and two peaks for the NP sample.

This chapter looks at some techniques of wax characterization and quantification; however, there are many other techniques that can be used and that present satisfactory results. The use of combined techniques may assist in the more accurate analysis of sample characteristics.

Acknowledgements

The authors thank Conselho Nacional de Pesquisa e Desenvolvimento (CNPq), Fundação Carlos Chagas de Amparo à Pesquisa do Estado do Rio de Janeiro (FAPERJ), and Petrobras for supporting this work.

Conflict of interest

The authors declare no competing financial interest.

Nomenclatures

API	American Petroleum Institute
ASTM	American Society for Testing and Materials
BF	brightfield
DSC	differential scanning calorimeter
GC	gas chromatography
HPLC	high performance liquid chromatography
NIR	near-infrared spectroscopy
NMR	nuclear magnetic resonance
NP	non-paraffinic
P_{1-4}	paraffinic petroleum
PL	polarized light
SAP	saturated, aromatic and polar
SARA	saturated, aromatic, resins and asphaltenes
TLC-FID	thin layer chromatography with flame ionization detection
UOP	universal oil products collection
WAT	wax appearance temperature
Q	total thermal effect of wax precipitation
c_w	wax precipitated concentration
$\overline{Q}$	constant thermal value of wax precipitation
T_f	final DSC temperature
w	mass content of precipitated waxes
ρ	specific mass
V_e	experimental volume used to the DSC measurement

Author details

Erika C.A. Nunes Chrisman[1]*, Angela C.P. Duncke[1], Márcia C.K. Oliveira[2]
and Márcio N. Souza[1]

1 Federal University of Rio de Janeiro, Rio de Janeiro, Brazil

2 Cenpes, Petrobras, Rio de Janeiro, Brazil

*Address all correspondence to: enunes@eq.ufrj.br

IntechOpen

References

[1] Farah MA. Petróleo e seus derivados: definição, constituição, aplicação, especificações, características de qualidade. Rio de Janeiro: LTC; 2012. 261 p

[2] Venkatesan R, Nagarajan NR, Paso K, Yi Y-B, Sastry AM, Fogler HS. The strength of paraffin gels formed under static and flow conditions. Chemical Engineering Science. 2005;**60**:3587-3598. DOI: 10.1016/j.ces.2005.02.045

[3] Hammami A, Raines MA. Paraffin deposition from crude oils: Comparison of laboratory results with field data. SPE Journal. 1999;**4**(1):9-18. DOI: 10.2118/54021-PA

[4] Kok MV, Varfolomeev MA, Nurgaliev DK. Wax appearance temperature (WAT) determinations of different origin crude oils by differential scanning calorimetry. Journal of Petroleum Science and Engineering. 2018;**168**:542-545. DOI: 10.1016/j.petrol.2018.05.045

[5] Lopes RT, Valente CM, De Jesus EFO, Camerini CS. Detection of paraffin deposition inside a draining tubulation by Compton scattering technique. Applied Radiation and Isotopes. 1997;**48**(10):1443-1450. DOI: 10.1016/S0969-8043(97)00255-8

[6] Azevedo LFA, Teixeira AM. A critical review of the modeling of wax deposition mechanisms. Petroleum Science and Technology. 2003;**21**(3–4):393-408. DOI: 10.1081/LFT-120018528

[7] Huang Z, Lee HS, Senra M, Fogler HS. A fundamental model of wax deposition in subsea oil pipelines. AICHE Journal. 2011;**57**(11):2955-2964. DOI: 10.1002/aic.12517

[8] Creek JL, Lund HJ, Brill JP, Volk M. Wax deposition in single phase flow. Fluid Phase Equilibria. 1999;**158**: 801-811. DOI: 10.1016/S0378-3812(99)00106-5

[9] Quan Q, Gong J, Wang W, Gao G. Study on the aging and critical carbon number of wax deposition with temperature for crude oils. Journal of Petroleum Science and Engineering. 2015;**130**:1-5. DOI: 10.1016/j.petrol.2015.03.026

[10] Zheng S, Fogler HS. Fundamental investigation of wax diffusion characteristics in water-in-oil emulsion. Industrial and Engineering Chemistry Research. 2015;**54**:4420-4428. DOI: 10.1021/ie501955e

[11] Lin M, Li C, Yang F, Ma Y. Isothermal structure development of Qinghai waxy crude oil after static and dynamic cooling. Journal of Petroleum Science and Engineering. 2011;**77**: 351-358. DOI: 10.1016/j.petrol.2011.04.010

[12] Eskin D, Ratulowski J, Akbarzadeh K. A model of wax deposit layer formation. Chemical Engineering Science. 2013;**97**:311-319. DOI: 10.1016/j.ces.2013.04.040

[13] Kasumu AS, Arumugam S, Mehrotra AK. Effect of cooling rate on the wax precipitation temperature of "waxy" mixtures. Fuel. 2013;**103**:1144-1147. DOI: 10.1016/j.fuel.2012.09.036

[14] Allen TO, Roberts AP. Production Operations: Well Completions, Workover, and Stimulation. 3rd ed. Vol. 2. Tulsa: Inc.; 1989. 364 p

[15] Srivastava SP, Tandon RS, Verma PS, Saxena AK, Joshi GC, Phatak SD. Crystallization behavior of n-paraffins in Bombay-High middle-distillate wax/gel. Fuel. 1992;**71**:533-537. DOI: 10.1016/0016-2361(92)90150-M

[16] Kané M, Djabourov M, Volle J-L, Lechaire J-P, Frebourg G. Morphology of paraffin crystals in waxy crude oils cooled in quiescent conditions and under flow. Fuel. 2003;**82**:127-135. DOI: 10.1016/S0016-2361(02)00222-3

[17] Hammami A, Mehrotra AK. Thermal behavior of polymorphic n-alkanes: effect of cooling rate on the major transition temperatures. Fuel. 1995;**74**:96-101. DOI: 10.1016/0016-2361(94)P4338-3

[18] Cazaux G, Barre L, Brucy F. Waxy crude cold start: Assessment through gel structural properties. In: SPE Annual Technical Conference and Exhibition, 27–30 September 1998; New Orleans. Louisiana: Society of Petroleum Engineers; 1998. pp. 729-739. DOI: 10.2118/49213-MS

[19] Jiang Z, Hutchinson JM, Imrie CT. Measurement of the wax appearance temperatures of crude oils by temperature modulated differential scanning calorimetry. Fuel. 2001;**80**: 367-371. DOI: 10.1016/S0016-2361(00)00092-2

[20] Guo X, Pethica BA, Huang JS, Adamson DH, Prud'homme RK. Effect of cooling rate on crystallization of model waxy oils with microcrystalline poly(ethylene butene). Energy & Fuels. 2006;**20**:250-256. DOI: 10.1021/ef050163e

[21] Paso K, Kallevik H, Sjoblom J. Measurement of wax appearance temperature using near-infrared (NIR) scattering. Energy & Fuels. 2009;**23**: 4988-4994. DOI: 10.1021/ef900173b

[22] Webber RMJ. Low temperature rheology of lubricating mineral oils: Effects of cooling rate and wax crystallization on flow properties of base oils. Journal of Rheology. 1999;**43**: 911-931. DOI: 10.1122/1.551045

[23] Meighani HM, Ghotbi C, Behbahani TJ, Sharifi K. A new investigation of wax precipitation in Iranian crude oils: Experimental method based on FTIR spectroscopy and theoretical predictions using PC-SAFT model. Journal of Molecular Liquids. 2018;**249**:970-979. DOI: 10.1016/j.molliq.2017.11.110

[24] Rønningsen HP, Bjoerndal B, Hansen AB, Pedersen WB. Wax precipitation from North Sea crude oils. 1. Crystallization and dissolution temperatures, and Newtonian and non-Newtonian flow properties. Energy & Fuels. 1991;**5**:895-908. DOI: 10.1021/ef00030a019

[25] Alcazar-Vara LA, Buenrostro-Gonzalez E. Characterization of the wax precipitation in Mexican crude oils. Fuel Processing Technology. 2011;**92**: 2366-2374. DOI: 10.1016/j.fuproc.2011.08.012

[26] Wessel R, Ball RC. Fractal aggregates and gels in shear flow. Physical Review A: Atomic, Molecular, and Optical Physics. 1992;**46**(6): 3008-3011. DOI: 10.1103/PhysRevA.46.R3008

[27] Soares EJ, Thompson RL, Machado A. Measuring the yielding of waxy crude oils considering its time-dependency and apparent-yield-stress nature. Applied Rheology. 2013;**23**:62798. DOI: 10.3933/ApplRheol-23-62798

[28] Andrade DEV, Da Cruz ACB, Franco AT, Negrão COR. Influence of the initial cooling temperature on the gelation and yield stress of waxy crude oils. Rheologica Acta. 2015;**54**:149-157. DOI: 10.1007/s00397-014-0812-0

[29] Létoffé JM, Claudy P, Kok MV, Garcin M, Volle JL. Crude oils: Characterization of waxes precipitated on cooling by d.s.c. and thermo microscopy. Fuel. 1995;**74**(6):810-817. DOI: 10.1016/0016-2361(94)00006-D

[30] Carlton RA. Polarized light. In: Carlton RA, editor. Pharmaceutical Microscopy. New York: Springer-Verlag New York; 2011. pp. 7-64. DOI: 10.1007/978-1-4419-8831-7

[31] Speight JG. The Chemistry and Technology of Petroleum. 3rd ed. New York: Marcel-Dekker; 1999. 918 p

[32] Pedersen KS, Rønningsen HP. Effect of precipitated wax on viscosity: A model for predicting non-Newtonian viscosity of crude oils. Energy & Fuels. 2000;14:43-51. DOI: 10.1021/ef9901185

[33] Li H, Zhang J. A generalized model for predicting non-Newtonian viscosity of waxy crude as a function of temperature and precipitated wax. Fuel. 2003;82:1387-1397. DOI: 10.1016/S0016-2361(03)00035-8

[34] Lee HS, Singh P, Thomason WH, Fogler HS. Waxy oil gel breaking mechanisms: Adhesive versus cohesive failure. Energy & Fuels. 2008;22(1): 480-487. DOI: 10.1021/ef700212v

[35] Chang C, Boger DV, Nguyen QD. Influence of thermal history on the waxy structure of statically cooled waxy crude oil. SPE Journal. 2000;5:148-157. DOI: 10.2118/57959-PA

[36] Bai C, Zhang J. Thermal, macroscopic and microscopic characteristics of wax deposits in fields pipelines. Energy & Fuels. 2013;27: 752-759

[37] Barbato C, Nogueira B, Khalil M, Fonseca R, Gonçalves M, Pinto JC, et al. Contribution to a more reproducible method for measuring yield stress of waxy crude oils emulsions. Energy & Fuels. 2014;28(3):1717-1725. DOI: 10.1021/ef401976r

[38] Silva JAL, Coutinho JAP. Dynamic rheological analysis of the gelation behavior of waxy crude oils. Rheologica Acta. 2004;43(5):433-441. DOI: 10.1007/s00397-004-0367-6

[39] Oliveira MCK, Texeira A, Vieira LC, Carvalho RM, Carvalho ABM, Couto BC. Flow assurance study for waxy crude oils. Energy & Fuels. 2012;26: 2688-2695. DOI: 10.1021/ef201407j

[40] Ruwoldt J, Kurniawan M, Oschmann H-J. Non-linear dependency of wax appearance temperature on cooling rate. Journal of Petroleum Science and Engineering. 2018;165: 114-126. DOI: 10.1016/j.petrol.2018.0 10.1016/j.petrol.2018.02.011 2.011

[41] Kok KM, Létoffé J-M, Claudy P, Martin D, Garcin M, Volle J-L. Comparison of wax appearance temperatures of crude oils by differential scanning calorimetry, thermomicroscopy and viscometry. Fuel. 1996;75(7):787-790. DOI: 10.1016/0016-2361(96)00046-4

[42] Srivastava SP, Handoo J, Agrawal KM, Joshi GC. Phase-transition studies in n-alkanes and petroleum-related waxes: A review. Journal of Physics and Chemistry of Solids. 1993;54:639-670. DOI: 10.1016/0022-3697(93)90126-C

[43] Guo X, Pethica BA, Huang JS, Prud'homme RK, Adamson DH, Fetters LJ. Crystallization of mixed paraffin from model waxy oils and the influence of micro-crystalline poly(ethylene-butene) random copolymers. Energy & Fuels. 2004;18(4):930-937. DOI: 10.1021/ef034098p

[44] Senra M, Panacharoensawad E, Kraiwattanawong K, Singh P, Fogler HS. Role of n-alkane polydispersity on the crystallization of n-alkanes from solution. Energy & Fuels. 2008;22: 545-555. DOI: 10.1021/ef700490k

[45] Senra M, Scholand T, Maxey C, Fogler HS. Role of polydispersity and co-crystallization on the gelation of long-chained n-alkanes in solution.

Energy & Fuels. 2009;**23**(12):5947-5957.
DOI: 10.1021/ef900652e

[46] Dirand M, Bouroukba M, Chevallier
V, Petitjean D, Behar E, Ruffier-Meray
V. Normal alkanes, multialkane
synthetic model mixtures, and real
petroleum waxes: Crystallographic
structures, thermodynamic properties,
and crystallization. Journal of Chemical
& Engineering Data. 2002;**47**(2):
115-143. DOI: 10.1021/je0100084

[47] Saxena H, Majhi A, Behera B.
Prediction of wax content in crude oil
and petroleum fraction by proton NMR.
Petroleum Science and Technology.
2018:1-8. DOI: 10.1080/
10916466.2018.1536713

[48] Yi S, Zhang J. Relationship between
waxy crude oil composition and change
in the morphology and structure of wax
crystals induced by pour-point-
depressant beneficiation. Energy &
Fuels. 2011;**25**:1686-1696. DOI: 10.1021/
ef200059p

[49] Chen J, Zhang J, Li H. Determining
the wax content of crude oils by using
differential scanning calorimetry.
Thermochimica Acta. 2004;**410**:23-26.
DOI: 10.1016/S0040-6031(03)00367-8

[50] Robustillo MD, Coto B, Martos C,
Espada JJ. Assessment of different
methods to determine the total wax
content of crude oils. Energy & Fuels.
2012;**26**:6352-6357. DOI: 10.1021/
ef301190s

Managing Paraffin/Wax Deposition Challenges in Deepwater Hydrocarbon Production Systems

Keshawa Shukla and Mayank Vishal Labh

Abstract

The prevention of solids formation and their deposition are major challenges to design and operate any subsea hydrocarbon production systems. One of the most challenging issues is the management of paraffin/wax. As the water depth increases, at the low temperatures of subsea conditions, hydrocarbons may precipitate as wax, which can solidify and restrict the flow. During shutdown of a subsea production system wax gel may form and solidify when a crude oil cools below its pour point causing operational problems from downhole to the processing facilities. The purpose of this chapter is to address the paraffin/wax formation and deposition issues to properly design a subsea production system consisting of pipe-in-pipe flowline and flexible riser under deepwater environment. A field specific example is presented to manage the wax formation/deposition and prevent paraffin/wax deposition risks in an effective way during the normal and the shut-in operations of the subsea production system. This study illustrates that the subsea hardware, such as water stop and equipment valves, along with the flowline, riser and jumper should be sufficiently insulated in order to prevent any cold spots in the production system, and achieve sufficient cooldown time for the shut-in operations.

Keywords: paraffin/wax, subsea production system, hydrocarbons, pipe-in-pipe flowline, flexible riser

1. Introduction

The major flow assurance challenges in the design and operation of a subsea hydrocarbon production system arise mainly due to the reservoir fluid properties, multiphase fluid flow, and solid formation such as paraffin/wax, hydrate, asphaltene, scale, corrosion, emulsion and foam. In particular, the formations of paraffin/wax and hydrate at low temperature and high pressure conditions in a deep water production system are critical to manage when transporting fluids from the reservoirs to the host facilities. The wax present in hydrocarbon fluids is mainly comprised of high molecular weight paraffinic compounds that are crystalline in nature. The wax can drop out of the crude oil at the wax appearance temperature (WAT) and deposit in the subsea systems during the production operations when the fluid temperature is lower than WAT. Below the pour point, the wax can gel and solidify resulting in restricting the flow and plugging the subsea system. Likewise, the hydrates can form and deposit in the subsea systems when the produced hydrocarbon gas and water mix at low temperature and high pressure (for example, see for review [1–6]).

The cooling down of a subsea production system in a shut-in process is another complex transient heat transfer problem. In this process, the fluid flow stops and heat transfer occurs between the subsea production system and the surrounding environment through the pipe wall, and the system eventually reaches to low ambient seawater temperatures. The rate at which the temperature drops with time becomes important to manage paraffin/wax deposition, hydrate formation and their solidifications in such subsea production systems.

When the operating temperatures are low, the cold spots can appear at inadequately insulated or uninsulated sections of the subsea structures and equipment including jumpers, flowlines, risers, manifolds, field joints, water stops, bulkheads and valves, among other components. Therefore, the subsea production systems should be sufficiently insulated for the wax and hydrate controls during both the normal operation and the shut-in operation. The shut-in operation normally requires a minimum cooldown time. Generally, the cooldown time is the period when the fluid temperature reaches the wax deposition temperature or hydrate formation temperature at the operating pressures during the shut-in operation. During this period the operator has to decide the remedial actions such as to commence chemical inhibition, depressurization and hot oil circulation to prevent plugging of the subsea production systems [1–6]. Note that in this study, the cooldown time is the time when the fluid temperature reaches the pour point in the shut-in operation to prevent wax gelling/solidification in any part of the production system.

In this chapter, a subsea production system is considered typical to the Gulf of Mexico (GoM). The production system consists of a pipe-in-pipe (PIP) flowline, a flexible riser, insulated jumpers, subsea structures and equipment. The flowline, riser and jumpers of this system are adequately insulated so that they can operate above the pour point and WAT during normal operations and provide a minimum cooldown time of 12 h to prevent any cold spots (low temperature conditions) and wax gelling/solidification during the shut-in operations.

The objective of this chapter is to investigate the cooldown time and cold spots (low temperature conditions) of the above assumed production system. The cold spots can arise due to uninsulated or inadequately insulated parts of the subsea structures and equipment, such as water stops and valves, during the shut-in operations.

The rest of the chapter is organized as follows. Section 2 describes the subsea field layout typical to the Gulf of Mexico (GoM). Section 3 describes the properties of the PIP flowline, dry and wet insulations for retaining heat in the subsea production system. This section also describes the design basis and the operating constraints. Section 4 presents the method and procedure employed to perform the cold spot analysis at different locations of the subsea production system. Section 5 presents simulation results for the cooldown time and cold spots of the subsea production system including subsea structures and equipment. Section 6 presents the conclusion of this study.

2. Subsea field description

Figure 1 shows the sketch of a subsea production system typical to the GoM. The system consists of a well with corresponding wellbore and wellhead (WH), four

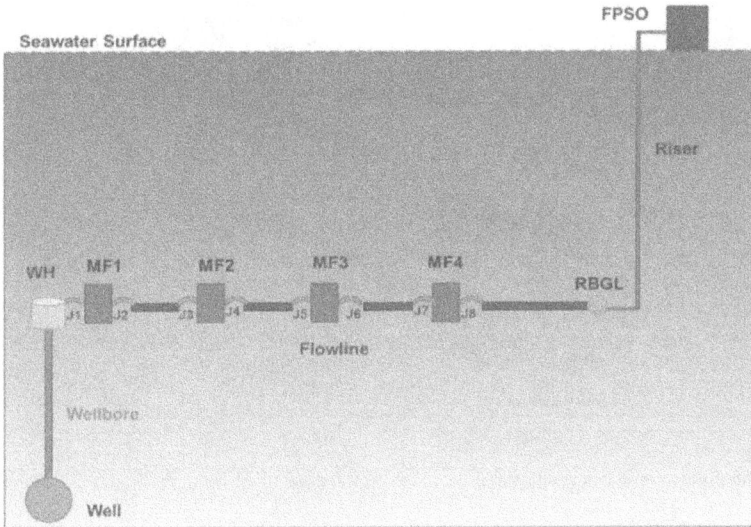

Figure 1.
A Schematic of the subsea production system.

manifolds (MF1, MF2, MF3 and MF4), and eight jumpers (J1–J8). The manifolds are connected to a flowline and a riser leading to a floating production storage and offloading (FPSO) facility. RBGL indicates the location of a riser base gas lift. Practically, the manifold MF1 has production from two wells, while each of MF2, MF3 and MF4 has production from a single well.

The total length of the flowline is approximately 17 km. The wellheads are located in water depth of about 1450 meter. The ambient seabed temperature is approximately 3.5°C. The flowline system consists of (2628.9 mm (8.625-inch) × 3886.2 mm (12.75-inch)) PIP flowline. The riser is a 2133.6 mm (7-inch) flexible riser starting from the riser base. The jumpers have 2628.9 mm (8.625-inch) stainless steel outside diameter (OD) with the glass syntactic polyurethane (GSPU) wet insulation. The subsea hardware such as the water stops is placed at the spacings of 700 meter, and equipment valves are added at the manifolds and RBGL.

3. Design basis and insulation materials

3.1 PIP flowline system and dry insulation

The PIP insulation is a passive non-chemical solution for flow assurance problems and does not need input of work and heat. Heat retention is achieved by surrounding the pipeline with materials that offer a high resistance to heat transfer with low thermal conductivity.

In a PIP system, a pipe is inserted inside another pipe. A dry insulation material, such as aerogel, is placed in the created intermediate annulus and is protected by the outer pipe from hydrostatic pressure and water penetration. Having a low thermal conductivity, aerogel allows the design of pipelines with the overall heat transfer coefficient (U_{ID}-value) significantly low without compromising the overall external dimensions of the PIP system. For the case of a rigid outer pipe, an air gap exists between the outside diameter surface of the insulation and inside diameter of the outer pipe adding to the heat resistance of the system. In the recent past, PIP flowline systems have been used for a number of deep water projects [1–2, 7–11].

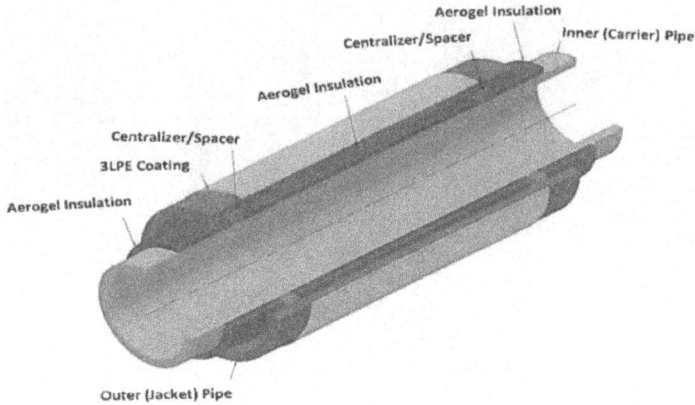

Figure 2.
Pipe-in-pipe insulation section of a pipe.

Figure 2 shows a typical PIP flowline section with the various layers of dry insulation [1].

In this study, the PIP insulation material is considered to be aerogel. The centralizer spacing at every stalk length is set about 2.2 m specified for aerogel thickness requirement for the reeled pipeline. The function of centralizer is to support the inner pipe centralized within the outer pipe to prevent possible damage to the PIP thermal insulation and transfer loads between the inner and outer pipes.

3.2 Wet insulation

Wet insulation does not require any input of energy such as work and heat. For example, glass syntactic polyurethane (GSPU) is the typical subsea wet insulation material. It can be used to retain heat in the jumpers and hardware providing U_{ID}-values greater than 1.0 W/m^2.K [1, 12]. The wet insulation is directly coated to steel pipes and placed on the seabed exposed to seawater.

In this study, the GSPU wet insulation is used along with fusion bonded epoxy (FBE) and three-layer polyethylene (3LPE) coatings for jumpers and equipment. Thermal conductivities of GSPU, FBE and 3LPE are 0.16, 0.3 and 0.4 W/m.K, respectively. **Figure 3** shows the schematic of a typical wet insulated pipe section [2].

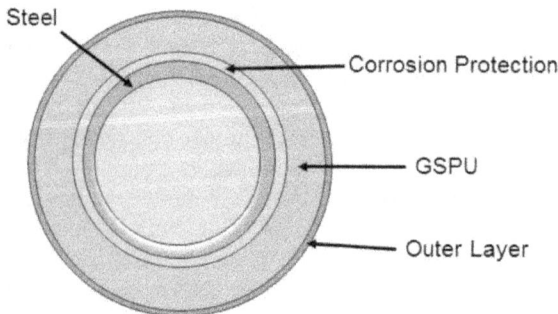

Figure 3.
A wet insulated pipe section.

3.3 Insulation thickness and cooldown time

Here, the PIP flowline insulation consists of 18.3 mm steel, 15 mm aerogel, 18.3 mm air, 19.1 mm steel and 3 mm 3LPE. The riser has the coating of 2133.6 mm flexible pipe. Jumpers are insulated with 1066.8 GSPU.

For the above flowline configuration and insulation, the U_{ID}-values of PIP flowline, flexible riser and wet insulated jumpers are 1.0, 3.5 and 2.9 $W/m^2.K$, respectively. These U_{ID}-values yield the required cooldown time of 12 h based on pour point of the waxy crude oil, that is, when the fluid temperature is equivalent to the wax pour point temperature during the shut-in operation.

The U_{ID}-values of water stops and valves are determined using their typical configurations and GSPU insulation as discussed below.

3.4 Crude oil properties

The crude oil comprises of waxy oil, gas and produced water with 33° API gravity. The wax appearance temperature and pour point/wax deposition temperature of the fluid are 29 and 18°C, respectively. The total liquid production is approximately 19,300 STBPD (stock tank barrel per day) with associated gas of 13 MMSCFD (million standard cubic feet per day), equally distributed to five wells. The watercut is approximately 10% by volume and gas to oil ratio (GOR) is 750 SCF/STB (standard cubic feet/standard barrel). The hydrate curve is determined from the fluid composition without any inhibitor. Note that the crude oil was characterized up to C_{80+} components and the PVT properties were determined using a multiphase software, PVTsim Nova 3. The composition of C_{18+} components was found to be greater than 11 mole%, indicating the presence of wax with wax content of 3–6 wt%.

3.5 Subsea system design constraints

For the assumed design and operation constraints of the jumpers, flowline and riser and their insulations, the required cooldown time should be 12 h for maintaining the fluid temperature above the wax gelling/solidification temperature, i.e., above the wax pour point temperature of 18°C. The normal arrival pressure and the ambient temperature at FPSO are set to be 19 bar and 19°C, respectively. Since the hydrate temperature of the fluid is always lower than the pour point, 12 h cooldown time is sufficient to manage the hydrate deposition during the shut-in operations.

4. Procedure of cold spot and cooldown time analyses

The cold spot and cooldown analyses were performed using the multiphase flow simulator OLGA 2016.2.1. The software uses a finite difference numerical scheme to solve mass, energy and momentum balances for multiphase fluid flow in a pipeline. The model accounts for the energy transfer between adjacent pipe segments, and inner pipe and surrounding. The fluid properties, hydrate dissociation curve and WAT were obtained from PVTsim Nova 3, which is a versatile equation of state modeling software.

The PIP flowline system was shut down via a linear ramp-down of the topsides valve on the facility and flow sources in the flowline closing simultaneously in 45 s. Steady state initial conditions were applied prior to the ramp-down. The cold spots were investigated at water stops and subsea equipment valves locations of the assumed subsea production system for several cases. However, the results for only two selected operating cases are presented below.

4.1 Case 1: flowline/riser/jumper system without equipment

Case 1 forms the base case providing sufficient insulations to PIP flowline, flexible riser and insulated jumper system without the water stops and equipment valves. In this case, the fluid temperature is always maintained above the wax appearance temperature (WAT) during the normal operations. In this case, the fluid temperature can be maintained above the pour point for up to 12 h (cooldown time) in the shut-in operations. Since the wax deposition issue dominates over the hydrate formation issue in this subsea production operation (that is the hydrate temperatures are always lower than 18°C pour point), only the pour point was utilized in determining the cooldown time.

4.2 Case 2: flowline/riser/jumper system with equipment (water stops and valves)

In this case, a typical equipment such as the water stop was added to the flowline to isolate a section of flooded annulus by preventing water passage to the adjacent PIP sections during installation and normal operations. The actual configuration of a water stop is shown in **Figure 4** [13]. A simplified water stop assembly used in this analysis is shown in **Figure 5**.

The two ends of the water stop consist of Hydrogenated Nitrile Butadiene Rubber (HNBR), which is the expected configuration so long as the middle steel section of the water stop does not touch the carrier pipe during normal operations. The middle part consists of the stainless steel with 3 mm coated 3LPE. This situation may occur if the middle steel section of the water stop touches the carrier pipe during normal operations. The HNBR material has good stability from thermal aging and is suitable for a water stop seal [9].

The water stops were placed at 700 meter intervals of PIP flowline assuming concentric layers surrounding the flowline with three segments of equal length 220 mm and thickness 33.3 mm (annular gap between inner and outer pipes). The

Figure 4.
Water stop configuration (TEKSEAL® Mechanical Clamp).

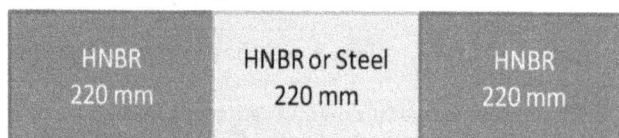

Figure 5.
A simplified water stop configuration.

Insulated (4.12 m)	Uninsulated (0.65 m)	Insulated (6.82 m)

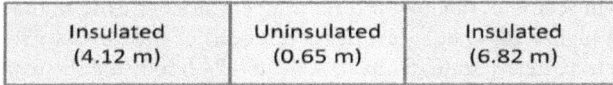

Figure 6.
A schematic of a valve insulation at RBGL manifold.

water stops were placed in the annulus of inner pipe and carrier pipe without any air gap. The thermal conductivity, specific heat and density of HNBR are 0.24 W/m.K, 0.25 J/kg.K and 1000 kg/m^3, respectively. In this study, a conservative case of the water stop configuration has been assumed.

In addition to the water stop, the typical subsea equipment valves were added at the manifolds and RBGL to assess the impact of cold spots on temperature. Such structures are commonly encountered in a subsea field development. The valves were insulated up to the bonnet but uninsulated on the actuator and pressure transmitters. The uninsulated valve section was accounted for by inserting a section of pipe with equivalent length of the valve bonnet diameter into the subsea hardware piping. The uninsulated subsea valve accumulators and pressure transmitters are modeled as cylindrical pipe segments. **Figure 6** shows the schematic of a valve insulation at the RBGL manifold. The similar configuration of equipment valves was assumed at other manifolds. All equipment valves were placed on the main flowline. The insulated sections of the valves used 1066.8 mm (3.5 inch) GSPU.

5. Results and discussions

This section summarizes results for the above two different operating scenarios.

5.1 Case 1: flowline/riser/jumper system without equipment

Figure 7 shows the fluid temperature variation with length of the flowline production system without any water stops and equipment valves. Also shown in the figure are WAT and pour point. The calculated U_{ID}-values of PIP flowline, flexible riser and wet insulated jumper are 1.0, 3.5 and 2.9 W/m^2.K, respectively.

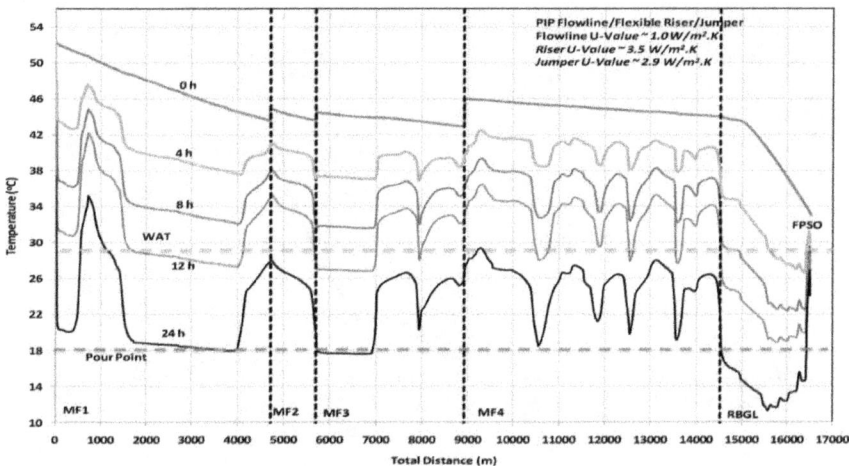

Figure 7.
Temperature vs. distance of flowline/riser/jumper system during shutdown.

For each shut-in scenario, the lowest fluid temperature lies close to the riser base because of the higher U_{ID}-value (less heat retention) of the flexible riser. Here, 0 h indicates results at steady state, while 4, 8, 12 and 24 h indicate the results for different shutdown times (hour).

Figure 8 shows pressure and temperature conditions for the normal and shut-in operations along with WAT, pour point and hydrate formation conditions. The results for the normal operation (0 h) show that the fluid temperature in the production system remains above WAT (29°C), pour point (18°C) and hydrate temperature. During shut-in operation, the fluid temperature remains above 18°C until 12 h shutdown time. However, the fluid cools below the wax pour point quickly after 12 h shutdown, and the fluid temperature lies in the wax gel region for 24 h shutdown.

The above results suggest that the combination of PIP flowline, flexible riser and wet insulated jumpers yields the cooldown time of 12 h, which can be sufficient to efficiently prevent cold spot (low temperature) problems arising from the wax gelling/solidification in the subsea production system.

5.2 Case 2: flowline/riser/jumper with equipment (water stops and valves)

In this case, the PIP flowline, flexible riser and wet insulated jumper system of Case 1 was assumed to include water stop seal assembly (HNBR with steel) and subsea equipment valves at manifolds and RBGL.

In order to check the overall performance of this system, it is first important to assess the impact of the assumed insulations on U_{ID}-values of water stop and equipment. **Figure 9** shows U_{ID}-values at water stops and equipment valves locations. For water stops the U_{ID}-value is greater than 50 W/m^2.K, and that for valves the U_{ID}-value is greater than 300 W/m^2.K. For such extremely large U_{ID}-values compared to those of Case 1 system, the cold spots can be expected at the water stops and equipment valves locations.

Figure 10 shows the temperature variation as a function of length of the production system with water stops and equipment valves for the normal and shut-in operation scenarios. The sections of the pipe near and at the location of water stops and equipment valves are seen to cool much faster than those of the flowline/riser/

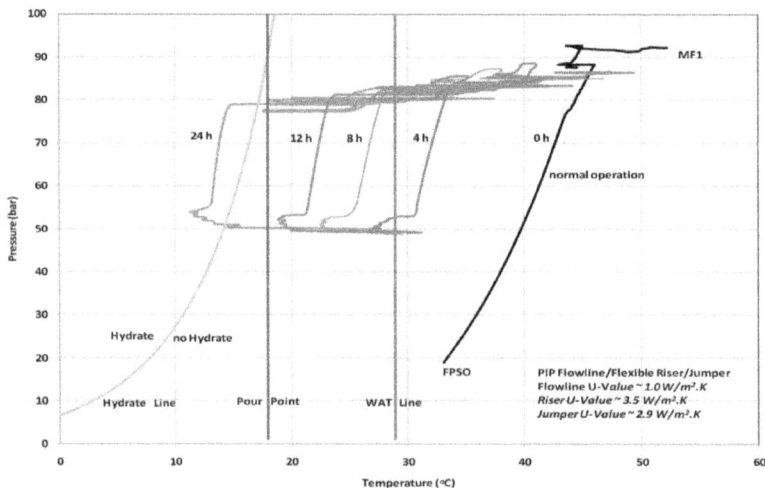

Figure 8.
Pressure vs. temperature of flowline/riser/jumper system during shutdown.

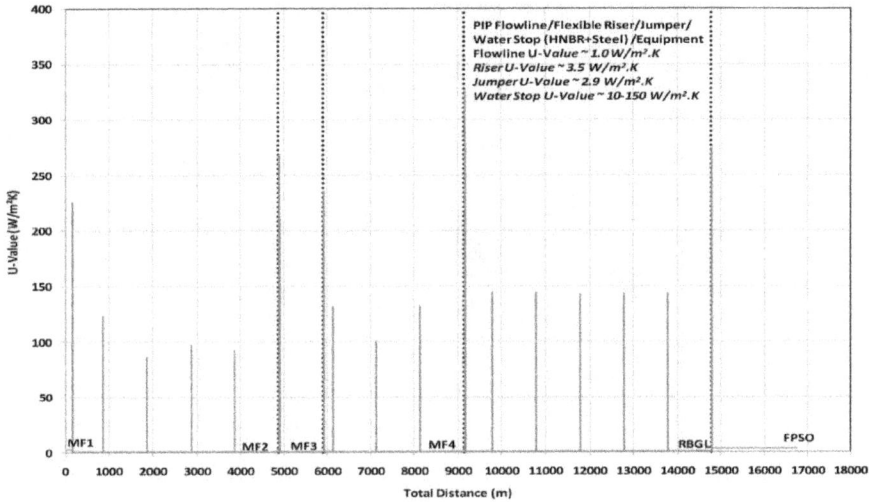

Figure 9.
U_{ID}-values for flowline/riser/jumper with water stops and valves.

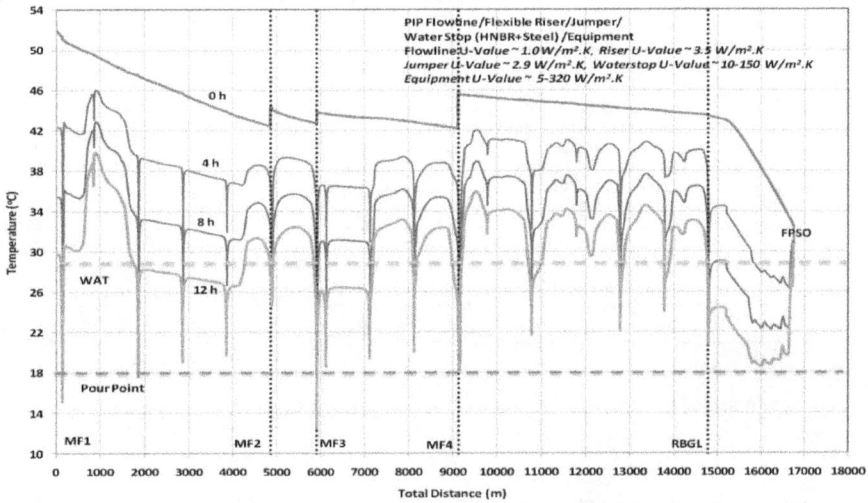

Figure 10.
Temperature vs. distance of flowline/riser/jumper system with water stops and equipment valves during shutdown.

jumper system. Due to the large U_{ID}-values of the stainless steel, the cooldown temperature at the water stops locations has lowered substantially (large downward spikes in temperature).

Figure 11 shows the variation of pressure with temperature during the shut-in operation. It shows the cooling temperature of the sections of flowline near and at locations of uninsulated and inadequately insulated equipment valves. The cold spots are not seen during normal operations because of the fact that only a small portion of the equipment is uninsulated, still maintaing the sufficient retention of heat. Because of the inadequate insulation of subsea water stops and equipment valves, however, the cold spots appear to yield only 8 h cooldown time, which is much less than the required 12 h cooldown time for shut-in operations. In this case,

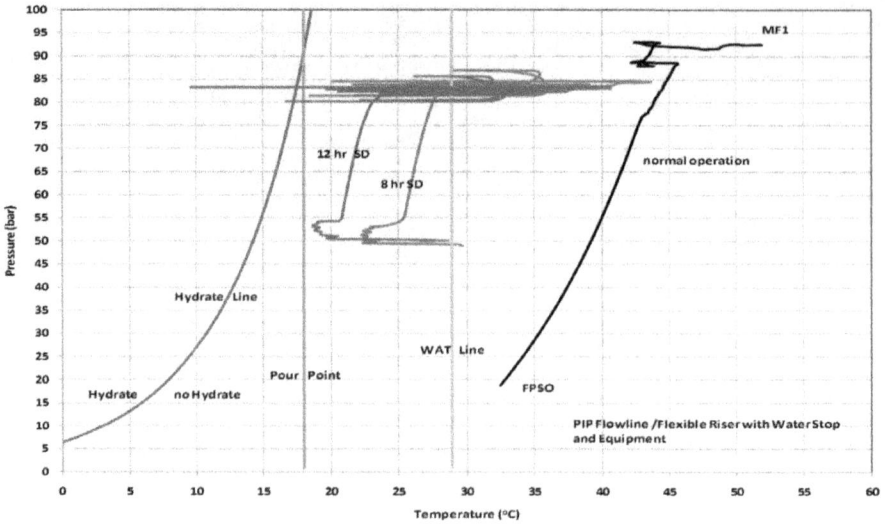

Figure 11.
Pressure vs. temperature of flowline/riser/jumper system with water stops and equipment valves during shutdown.

Shut-in operation	Cooldown temperature (°C)	
Time (h)	Case 1: flowline, riser and jumpers	Case 2: flowline, riser and jumpers with water stops/equipment valves
0	43	42
4	37	23
8	32	19
12	27	15
24	18	11

Table 1.
Cooling down time and temperature for Case 1 and Case 2 operating scenarios.

8 h cooldown is not sufficient to take remedial actions, especially for unplanned shutdowns. However, if feasible to insulate the entire equipment valve system with 1066.8 mm (3.5 inch) GSPU (without causing any installation and operation issue), it could provide the required cooldown time of 12 h.

Table 1 shows the summary of cooldown time achieved for Case 1 and Case 2 operating scenarios. As the table shows, Case 2 system cools down to 15°C after 12 h cooling time and cannot meet the cooldown time requirement of 12 h for the shut-in operations.

6. Conclusions

The assumed PIP flowline/flexible riser/wet insulated jumper system of the base case provides sufficient insulation for maintaining the fluid temperature above the wax pour point and hydrate deposition temperatures. This system could achieve the required cooldown time of 12 h, which is sufficient to keep the production system out of the wax gel formation region and avoid any cold spot (low temperature) in the shut-in operations.

When the water stops (HNBR + Steel) and partly insulated equipment valves were added to the PIP flowline/flexible riser/wet insulated jumper system, there appears no major issue of the cold spots during the normal operations. However, for the shut-in operations, the system shows cold spots (low temperature conditions) at the hardware (water stop and equipment valves) locations and can barely yield the cooldown time of 8 h. These results suggest that the uninsulated section of the equipment valves at manifolds should be adequately insulated in order to prevent any cold spots in the production system during the shut-in operations, even though the flowline, riser and jumpers are sufficiently insulated.

It is recommended to insulate the subsea hardware as much as possible. If the subsea structures and equipment cannot be sufficiently insulated due to installation and/or any other manufacturing reasons, it is recommended to manage the shut-in operations in 8 h and take preventive measures for wax gel formation/solidification. Typical actions to control wax deposition/solidification can be to maintain high operating temperature, inject chemical inhibitors, circulate hot oil and prepare for the pigging of the subsea production system. Such actions along with the recent subsea processing technologies can help reduce both capital and operating costs significantly, especially during the shut-in operations.

In the future study, the sensitivity analysis will be performed using different types of the crude oils with varying ratios of paraffinic hydrocarbons relevant to the deep water production systems.

Acknowledgements

Keshawa Shukla would like to thank the Subsea Engineering Program under EASA for investing his time and effort and Petroleum Engineering Department for giving access to the multiphase flow simulators at College of Engineering, Texas A&M University, College Station to contribute this chapter to the book which are gratefully acknowledged. Mayank Labh is a graduate student in Subsea Engineering (EASA) and has partly contributed to this work during his Directed Study. The open access publishing fees for this article have been covered by the Texas A&M University Open Access to Knowledge Fund (OAKFund), supported by the University Libraries and the Office of the Vice President for Research.

Conflict of interest

No potential conflict of interest.

Author details

Keshawa Shukla* and Mayank Vishal Labh
Subsea Engineering, EASA, College of Engineering, Texas A&M University,
College Station, TX, USA

*Address all correspondence to: kpshukla@tamu.edu

IntechOpen

References

[1] Shukla K. Non-chemical products offer effective flow assurance solutions. Offshore. April Issue. 2014:120

[2] Chapman M, Shukla K. Non-chemical solutions enhance flow assurance options. Offshore. April Issue. 2012:113

[3] Al-Safran EM, Brill JP. Applied Multiphase Flow in Pipes and Flow Assurance. Oil and Gas Production. Richardson, TX, USA: Society of Petroleum Engineers Press; 2017

[4] Huang Z, Zheng S, Fogler HC. Wax deposition—Experimental characterization. In: Theoretical Modeling and Field Practices. New York: CRC Press, Taylor and Francis Group; 2015

[5] Firoozabadi, A. Thermodynamics and applications in hydrocarbon energy production. New York: McGraw-Hill Education; 2016

[6] Gudmundsson JS. Flow Assurance Solids in Oil and Gas Production-Oil and Gas Production. London: CRC Press, Taylor and Francis Group; 2018

[7] de Azevedo FB, Solano RF, Manouchehri S, Dolinski A, Denniel S. Design, Fabrication and Installation of the First Ever reeled Pipe-in-Pipe System in Offshore Brazil. Offshore Technology Conference, May 4–7, Houston, TX, USA. 2009. p. 19951

[8] Decrin MK, Nebell F, de Naurois H, Parenteau T. Flow Assurance Modeling using an Electrical Trace Heated Pipe-in-Pipe: From Qualification to Offshore Testing. Offshore Technology Conference. May 6–9, Houston, TX, USA; 2013. p. 24060

[9] Jukes P, Delille F, Harrison G. Deepwater Pipe-in-Pipe (PIP) Qualification Testing for 350°F Service.

In: Proceedings of the 3rd International Offshore Pipeline Forum (IOPF), Houston, TX, October 29–30, 2008. p. 922

[10] Rao V, Mair J, Sriskandarajah T, Jones R, Booth P. Pipe-in-Pipe Swaged Field Joint for Reel Lay. Offshore Technology Conference. May 6–9, Houston, TX, USA; 2013. p. 24077

[11] Dixon M. Pipe-in-Pipe: Thermal Management for Effective Flow Assurance. Offshore Technology Conference. May 6–9, Houston, TX, USA; 2013. p. 24122

[12] Ruschau GR, Rogers RH, Woodley SA, Wright EJ. Evaluation and Qualification of Materials for Subsea Insulation Systems. SPE Annual Technical Conference and Exhibition, Sep 20–22, Florence, Italy; 2010. p. 131618

[13] TEKSEAL® Mechanical Clamp (Courtesy of Tekmare, Private Communication, 2016)

Chapter 4

Solar Thermal Energy Storage Using Paraffins as Phase Change Materials for Air Conditioning in the Built Environment

Wenye Lin, Zhenjun Ma, Haoshan Ren, Jingjing Liu and Kehua Li

Abstract

Thermal energy storage (TES) using phase change materials (PCMs) has received increasing attention since the last decades, due to its great potential for energy savings and energy management in the building sector. As one of the main categories of organic PCMs, paraffins exhibit favourable phase change temperatures for solar thermal energy storage. Its application is therefore effective to overcome the intermittent problem of solar energy utilisation, thereby reducing the power consumption of heating, ventilation and air conditioning (HVAC) systems and domestic hot water (DHW) systems. This chapter reviews the development and performance evaluation of solar thermal energy storage using paraffin-based PCMs in the built environment. Two case studies of solar-assisted radiant heating and desiccant cooling systems with integrated paraffin-based PCM TES were also presented. The results showed that paraffin-based PCM TES systems can rationalise the utilisation of solar thermal energy for air conditioning while maintaining a comfortable indoor environment.

Keywords: thermal energy storage, phase change materials, HVAC systems, solar energy, built environment

1. Introduction

As one of the major energy consumers, buildings account for around 45% of the global energy consumption with a similar share of greenhouse gases emissions [1]. Due to population increase, urbanisation, economic growth and improvement in the quality of life, energy usage in the building sector continues to rise. A study from the International Energy Agency [2] showed that without action, the energy demand in buildings could increase by 30% by 2060. A significant proportion of the energy demand from buildings is for building services, including heating, ventilation and air conditioning (HVAC) and domestic hot water (DHW) [3], in which the energy demand for HVAC is projected to increase by more than 70% from 2010 to 2050 [4]. Since the recent decades, the integration of renewable energies has been widely recognised as one of the effective solutions to reduce the HVAC power consumption in buildings, especially the utilisation of solar thermal energy. As one of the most attractive renewable energies, solar thermal energy is not only an ideal heat source for direct indoor space heating but also can be used to provide renewable cooling (e.g. absorption/adsorption cooling). However, due to the fact that solar energy is intermittent, the integration of solar thermal systems with thermal energy storage (TES) is therefore essential to rationalising energy management [5]. Among various TES technologies, TES using phase change materials (PCMs) has been receiving increasing attention. PCMs are substances that can absorb, store and release a large amount of thermal energy within a narrow temperature range through phase transitions [6], in which solid–liquid PCMs with substantial alternatives and a small change in volume during the phase change process are well suited for TES applications in the built environment [7]. Compared to sensible heat storage, TES using PCMs not only shows a significant reduction in the storage volume [8] but also enables the use of thermal energy at relatively constant temperatures [9].

PCMs are mainly categorised as organic, inorganic and eutectic materials, in which organic PCMs can be further classified as paraffins and non-paraffins [10], as shown in **Figure 1**. As PCMs, paraffins have a wide range of phase change temperatures [11], covering the temperature range from subzero to over 100°C [12]. Besides the desired phase change temperature ranges, paraffins have the advantages of congruent phase transition, self-nucleation to avoid supercooling, non-corrosiveness, long-term chemical stability without segregation and commercial availability at reasonable costs [13, 14]. However, paraffins have flammability, low thermal conductivity and relatively low volumetric latent heat storage density [15, 16].

The favourable phase change temperatures of the paraffins with phase transition temperatures at around and above 60°C, together with the other aforementioned advantages, make it one of the desired candidates for solar TES in the built environment to facilitate the solar-assisted HVAC and DHW generation. This chapter mainly focuses on solar TES using paraffin-based PCMs (with phase change temperature of and higher than 60°C) to facilitate the indoor air conditioning in the built environment. This chapter is structured as follows: Section 2 provides an overview of the solar TES using paraffin-based PCMs which can be used to facilitate the

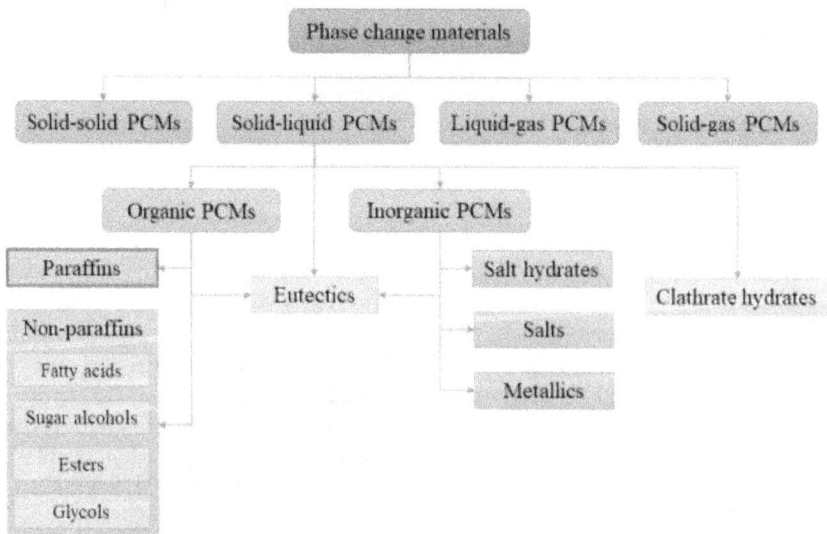

Figure 1.
PCM classifications.

indoor air conditioning. Sections 3 and 4 present two case studies of solar-assisted radiant space heating and desiccant cooling systems with paraffin-based PCMs, respectively. Section 5 provides a summary of this chapter.

2. Overview of thermal energy storage using paraffin-based PCMs in buildings

There are two main popular approaches to utilising paraffins as PCMs in the built environment. Paraffin-based PCMs can be integrated with solar thermal collectors to improve the system thermal efficiency, meanwhile serving as on-site TES. Alternatively, they can be used as independent TES units coupling with solar thermal collectors to provide continuous heat supply for the demand side. In both approaches, the charging of paraffins with the heat generated needs to be fulfilled first, followed by the retrieval of the heat using heat transfer fluids (HTFs) for specific applications (e.g. space heating or cooling). Accordingly, the following review is mainly segmented into two subsections based on the two stages. The utilisation of paraffin-based PCM TES in different solar hot water systems was also discussed and included in the first subsection, since there is a potential utilisation of the hot water generated to drive air conditioning systems. The paraffin-based PCMs used for TES in the built environment in this overview are summarised in **Table 1**.

2.1 Solar thermal energy storage using paraffin-based PCMs

2.1.1 Integration of paraffin-based PCMs with solar thermal collectors

Integrating PCM with solar collectors can not only reduce the highest temperature of the solar collectors, thereby extending the lifetime [17] and increasing the system thermal efficiency [18], but also fulfil on-site thermal storage [19]. For instance, a paraffin with a phase change temperature of around 60°C was enhanced using nano-Cu additives and laminated in a flat plate solar collector by Al-Kayiem and Lin [20] for water heating application. The experimental study showed that considerable thermal efficiency improvement was achieved with integrating the paraffin in the solar collector; however, the enhancement in thermal conductivity using nano-Cu particles showed limited benefits. A number of PCM/compressed expanded natural graphite (CENG) composites were prepared and integrated beneath a flat plate solar water heater by Haillot et al. [19, 21] for thermal performance enhancement. The characterisation of a number of PCM candidates demonstrated that the paraffin-based PCM composite, i.e. RT65/CENG, was the most suitable material to be used, due to its high thermal stability, conductivity and storage density. It was found that the solar fraction of the system using RT65/CENG composite can be effectively enhanced in summer; however, a low solar fraction was found in winter due to the high heat loss of the flat plate solar collectors.

With respect to the low heat loss, the integration of paraffin-based PCMs with evacuated tube collectors seems to be more promising. For instance, a paraffin wax with a melting temperature of 67°C was filled in the manifold of evacuated tube heat pipe solar collectors as a PCM TES unit by Naghavi et al. [22] to improve the performance of hot water supply. The numerical study demonstrated that the proposed system with PCM can maintain a high thermal efficiency of 55–60% which was less sensitive to the change of the draw-off water flowrate, compared to a conventional DHW system without PCM TES. Tritriacontane (i.e. $C_{33}H_{68}$) and erythritol were integrated into evacuated tubes simultaneously by Papadimitratos et al. [23] to gain the functionality of thermal storage while enhancing the system

Index	PCM	Phase change temperature	Application location	Application	Ref.
1	RT65	55–66°C	Solar collector—flat plate	Water heating	[19]
2	Paraffin	58.7–60.5°C	Solar collector—flat plate	Water heating	[20]
3	Paraffin	64°C	Solar collector—evacuated tubes	Water heating	[22]
4	Tritriacontane	72°C	Solar collector—evacuated tubes	Water heating	[23]
5	Paraffin	58–62°C	Solar collector—evacuated tubes	Water heating	[24, 25]
6	Paraffin	60°C	TES unit—packed bed and HTF tank	Water heating	[28]
7	Paraffin	62°C	TES unit—packed bed and HTF tank	Water heating	[29]
8	Paraffin	60 ± 2°C	TES unit—HTF tank	Water heating	[30]
9	Paraffin	55–62°C	TES unit—HTF tank	Water heating	[31]
10	Paraffin	60–62°C	TES unit—packed bed and heat exchanger	Water heating	[32]
11	Paraffin	56.06–64.99°C	TES unit—heat exchanger	Water heating	[33]
12	Paraffin	60°C	TES unit—heat exchanger	Air heating	[34]
13	RT65	55–66°C	TES unit—packed bed	Water heating	[21]
14	RT60	55–61°C	TES unit—heat exchanger	Solid desiccant cooling	[35]
15	RT65	57–68°C	TES unit—heat exchanger	Solid desiccant cooling	[35]
16	RT70HC	69–71°C	TES unit—heat exchanger	Solid desiccant cooling	[35]
17	Paraffin	67.2°C (optimal value)	TES unit—heat exchanger	Solid desiccant cooling	[36]
18	RT82	77–85°C	TES unit—heat exchanger	Liquid desiccant cooling	[37, 39]
19	RT100	99°C	TES unit—heat exchanger	Liquid desiccant cooling	[40]
20	Paraffin	6–62°C	Building envelopes	Floor radiant heating	[41]

Table 1.
Summary of paraffins used as PCMs for TES in the built environment.

thermal efficiency. A series of experiments were carried out based on the PCM-enhanced solar water heaters. The results showed that the evacuated tubes with integrated paraffin (i.e. tritriacontane) outperformed the ones with erythritol under a normal operation mode with continuous water circulation, due to its proper phase change temperature at around 72°C. It was also found that the thermal efficiency was improved 26% under the normal operation by using both PCMs simultaneously, compared to a traditional solar water heating (SWH) without using

PCMs. A paraffin wax with the melting temperature of 58–62°C was used as PCM and filled into evacuated tubes for thermal energy storage by Abokersh et al. [24]. The heat transfer between the water and PCM was achieved by different U-tube heat exchangers with and without fins inside the evacuated tubes, respectively. The experimental tests showed that the total energy efficiency can be improved by 35.8 and 47.7% for the PCM-enhanced evacuated tubes with and without fins, respectively, compared to a traditional forced recirculation SWH system. The further study [25] found that even the use of fins hindered the convective heat transfer within the molten PCM during the charging process, and its substantial contribution to the heat transfer enhancement during the PCM discharging process benefited the overall energy efficiency of the system.

2.1.2 Using paraffin-based PCMs as TES units

When PCM was used independent from solar thermal collectors, one of the scenarios is to install the PCM TES component in the heat transfer fluid tanks to fulfil hybrid sensible and latent heat storage. In this scenario, besides increasing the TES capacity, the paraffin-based PCMs also play the role in enhancing the thermal stratification for the water in the tanks [26], which relieves the loss caused by direct mixing of cold water with hot water. The selection of PCMs with proper phase change temperature and confinement geometry was reported to be significant [27]. For instance, an encapsulated PCM was packed in a water tank as a combined sensible and latent heat TES unit by [28] for DHW application. The PCM used is a paraffin (with a melting temperature of 60°C) encapsulated in spherical capsules. Two types of discharging experiments with continuous and batch-wise hot water retrieval processes were carried out, from which it was found that the batch-wise discharging best suited for the applications with intermittent hot water demands. A similar PCM TES packed bed with a paraffin (with a melting temperature of around 62°C) encapsulated in spherical capsules was tested by Ledesma et al. [29] for a SWH system. The numerical thermal performance analysis indicated the importance of system matching when coupled with the PCM TES unit and the SWH system whose outlet water temperature needs to be high enough for PCM charging. A paraffin encapsulated in aluminium cylinders was used as the heat storage media by Padmaraju et al. [30] for a DHW system. The comparative test results showed that the thermal energy stored in the paraffin-based PCM TES system far exceeded that stored in a sensible heat storage system of the same size of the storage tank. A similar conclusion was resulted by Kanimozhi and Bapu [31] through an experimental test based on a TES system with a paraffin filled in a number of copper tubes.

Different from the first scenario, the second scenario utilised the PCM TES units as heat exchangers for latent heat storage only. In this scenario, the higher heat transfer effectiveness is one of the keys to focus. For instance, a water-based multi-PCM pack bed TES unit for solar heat storage was numerically investigated by Aldoss and Rahman [32], in which three types of paraffins with different phase change temperatures were encapsulated in spherical capsules and placed at different sections of the TES unit serving as different thermal energy storage stages. It was found that the multi-PCM design can improve the system dynamic performance by increasing the charging and discharging rates. However, only limited thermal benefit can be achieved by further increasing the stage number. A paraffin wax (with the melting temperature of around 56–65°C) was pulled into the cell side of a shell and tube heat exchanger by Mahfuz et al. [33] for thermal energy storage in a SWH system. The energy, exergy and life cycle cost of the system were analysed experimentally under various flow rates. It was found that a higher flow rate was beneficial to gaining a higher energy efficiency and a lower life cycle cost, while it resulted in a lower exergy efficiency. An air-based PCM packed bed was tested by Karthikeyan and Velraj [34] to validate a number of latent TES packed bed models. The experimental measurement was used to identify the suitable models for PCM TES packed bed units when using different working fluids as the HTFs.

2.2 Paraffin-based PCM-assisted HVAC systems

After charged with thermal energy, the paraffin-based PCMs can be utilised to facilitate the indoor space heating directly or for indoor space cooling with the assistance of desiccant cooling devices. Either air or water can be used as the HTF in the systems, depending on the regeneration requirements. For instance, an air-based PCM TES unit was coupled with a solar-powered rotary desiccant cooling system by Ren et al. [35] to overcome the mismatch between energy demand for desiccant wheel regeneration and thermal energy generation from a hybrid photovoltaic thermal collector-solar air heater (PVT-SAH). The feasibility of using four paraffin-based PCMs (i.e. RT55, RT60, RT65 and RT70HC) as the TES media was investigated numerically in the proposed system. The results identified a near optimal system design for individual scenarios, in which RT65 was found to be the optimal paraffin-based PCM. When increasing the regeneration temperature from 60 to 70°C, the unsatisfied factor for supply air humidity ratio can be reduced from 24.2 to 6.0%, despite that it reduced the solar thermal contribution from 100.0 to 82.6%. The PVT-SAH and PCM-assisted rotary desiccant cooling systems were then further optimised to maximise its energy performance by the same authors [36] using a multilayer perceptron neural network and a genetic algorithm. It was found that the PCM phase change temperature was one of the most important factors, whose optimal value was 67.2°C. The design optimisation identified an optimal design; by using which, the specific net power generation and the solar thermal contribution of the proposed system can reach 10.32 kWh/m^2 and 99.4%, respectively, compared to that of 3.77 kWh/m^2 and 91.5% for a baseline case without optimisation. These studies indicated the importance of using the paraffin with proper thermal properties and optimal coupling of PCM TES in a solar-assisted desiccant cooling system for performance improvement.

Besides solid desiccant cooling, paraffin-based PCM TES designed for the regeneration of liquid desiccant materials was also reported. For instance, a triplex tube heat exchanger with integrated PCM as a TES unit was developed by Al-Abidi et al. [37, 38] and Mat et al. [39] for liquid desiccant air conditioning systems. A series of numerical modelling and experimental studies were carried out to investigate the thermal performance of the PCM TES unit. The results showed that the phase change time required can be reduced by more than 50%, if the triplex tube was intensively finned both internally and externally, and the melting process of the PCM can be accelerated by heating on both sides of the triplex tube. PCM TES units with various heat transfer enhancement techniques, including circular fins, longitudinal fins and multi-tube systems, were developed and experimentally investigated by Agyenim [40] to facilitate solar power absorption cooling systems and space heating/hot water systems. It was found that the multi-tube and longitudinal finned PCM TES units presented the most favourable charging and discharging performance, whose overall thermal energy utilisation efficiency reached 83.2% and 82.0%, respectively. It was therefore recommended to combine two heat transfer enhancement techniques to optimise the thermal performance of the PCM TES unit.

It is worthwhile to mention that another potential application of paraffins is to integrate paraffin-based PCMs into building envelopes for demand side management. For instance, a number of shape-stabilised PCMs were prepared by Zhang et al. [41], in which the ones with the melting temperature of 60–62°C were developed for the electric underfloor space heating system, thereby facilitating the peak-load shifting and making use of the electricity tariff. The authors highlighted that building energy efficiency can be significantly improved by combining radiant floor heating and thermal storage. Even though the PCM layer reported in this study used electrical heat as the heat source, it can be easily modified by integrating with hot water/air hydraulic piping/ducting to store and distribute the solar heat.

3. Case study I: solar-assisted heating system with integrated paraffin-based PCMs

The rationalisation of solar thermal energy utilisation is an alternative solution to facilitate indoor space heating. **Figure 2** illustrates the schematic of a solar-assisted radiant heating system with integrated paraffin-based PCM TES. It mainly consists of evacuated tube solar collectors, a paraffin-based PCM TES unit, two pumps, an auxiliary electric heater, the terminal heat-distributing devices which are radiant floor panels in this study and the corresponding piping system. In this system, the evacuated tube solar collectors were used to generate hot water, which can then be supplied for indoor space heating directly through the radiant floor heating panels, or used to charge the PCM TES unit, or both, during the daytime. During the night-time, the indoor space heating was achieved by circulating the water between the PCM TES unit and the radiant floor heating panels to retrieve the stored heat for indoor space heating. It is worthwhile to mention that the discharging water flow directed through the PCM TES is reversed compared to the charging water flow, so as to maximise the thermal performance of the PCM TES unit. The supply water temperature for the radiant floor panels was controlled to be constant by mixing a fraction of the return water with the hot water supplied from the evacuated tube or the PCM TES unit. The auxiliary electric heater can be used to maintain the desired supply water temperature when the thermal energy generated or stored is not sufficient. The indoor heating demand was satisfied by varying the hot water flow rate through the radiant floor panels through changing the operating speed of the supply water pump.

The system performance was evaluated numerically using TRNSYS simulation studio [42]. In the system modelling, the building heating load of a typical Australian house with an air-conditioned floor area of 150 m² [43, 44] under Sydney winter

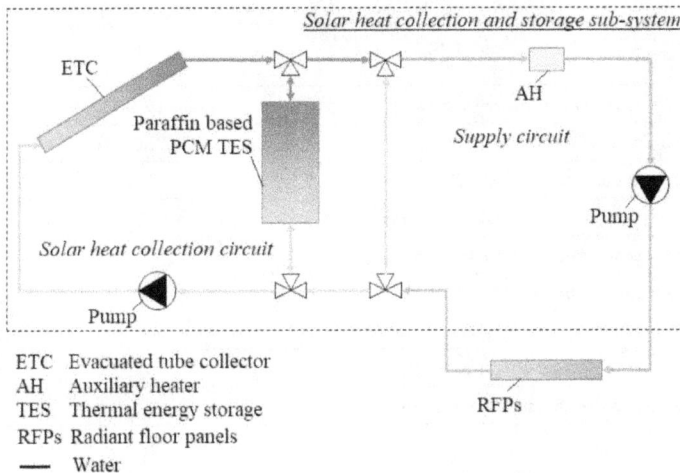

Figure 2.
Schematic of the solar-assisted radiant heating system with integrated paraffin-based PCM TES.

weather condition was modelled and used as the heating demand to be covered by the proposed system. This building heating load was simulated using Type 56 in TRNSYS based on the indoor air temperature setting of 20°C and the internal loads, occupancy schedule and internal adjustable shading settings required by the Australian Nationwide House Energy Rating Scheme (NatHERS) [45]. The evacuated tube solar collector, the auxiliary electric heater and the pumps employed were modelled using Type 71, Type 6 and Type 3 in TRNSYS, respectively. The radiant floor heating panels were modelled using an upgraded Type 1231 which was slightly revised by replacing the mean temperature difference with the log mean temperature difference to improve its accuracy. The PCM TES unit was a water-based tube-in-tank heat exchanger, in which the paraffin was encapsulated in the tube-side with water flowing through the cylinder-side. The PCM TES model was developed using an enhanced enthalpy method for accurate modelling of the phase change process and the finite difference method for discretisation of the energy balance equations. A similar PCM TES model can be found in Bourne and Novoselac [46]. The paraffin-based PCM used is a commercial PCM product RT69HC from Rubitherm [47], with a nominal phase change temperature of around 69°C. The key parameters used in the numerical system performance evaluation are summarised in **Table 2**.

Figure 3 presents the performance of the solar-assisted radiant heating system with the paraffin-based PCM over 3 winter days (note that the simulation results over an additional day before the 3 test days were not reported to avoid the influence from initial values). It can be seen from **Figure 3a** that the solar thermal energy collected and stored can fully cover the heating demand. The pumps were the only power consumers, in which the pump in the solar heat collection circuit was turned on during the daytime when the solar energy was sufficient to heat the water, while the power consumption of the pump in the supply circuit seemed to present a proportion trend to the heating load. Total power consumption was only 0.52 kWh which was much lower than the heating demand of 115.33 kWh over the 3 test days. **Figure 3b** illustrates the temperature variation of the inlet and outlet water of the paraffin-based PCM TES unit. When the hot water from the evacuated tube solar collector was drawn for PCM charging (highlighted with the red background), a clear thermal charging process can be observed, which presented a relatively constant outlet water temperature from the PCM TES unit. During the PCM discharging period, due to the reversed water flow through the PCM TES unit, a high outlet

Parameter	Radiant heating	Desiccant cooling
Area of the evacuated tube solar collector (m²)	26.24	59.04
Type of paraffin-based PCM	Rt69HC [47]	RT69HC [47]
Total amount of the paraffin-based PCM (kg)	632.7	1476.3
Power of the pump in the solar heat collection circuit (W)	15	38
Maximal power of the pump in the supply circuit (W)	35	80
Supply water temperature setting (°C)	60	64
Maximal power of the supply fan (W)	—	533.3
Maximal power of the regeneration fan (W)	—	533.3
Desiccant wheel outlet air humidity setting (g/kg)	—	8.1

Table 2.
Key parameters used in the performance evaluation of the solar-assisted radiant heating and desiccant cooling systems with integrated paraffin-based PCM TES.

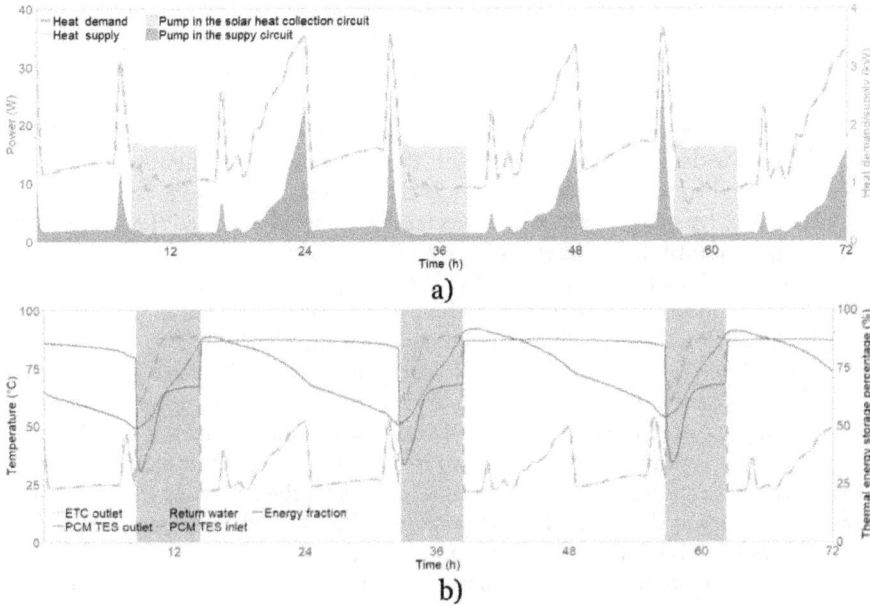

Figure 3.
*Modelling results for the solar-assisted radiant heating system with integrated paraffin-based PCM TES.
(a) Power consumption and heating energy demand. (b) Inlet and outlet water temperatures of the paraffin-
based PCM TES unit.*

water temperature from the PCM TES unit was achieved. It enabled the supply of
a high-temperature water for space heating, even though the return water from the
radiant floor heating panels was low. Correspondingly, the thermal energy storage
percentage in the paraffin-based PCM increased during the PCM charging periods
rapidly and then reduced during the PCM discharging periods gradually, which
varied from 48.96 to 91.54% over the 3 test winter days.

4. Case study II: solar-assisted cooling system with integrated paraffin-based PCMs

Rotary desiccant cooling systems, which combine rotary desiccant dehumidification and evaporative cooling technologies, have been recognised as an alternative
to conventional vapour compression air conditioning systems [48, 49]. It offers the
advantages including being free from CFCs, using low-grade thermal energy, and
independent humidity and temperature control, which therefore is more energy
efficient and environmentally friendly than conventional vapour compression air
conditioning systems [49]. In a rotary desiccant cooling system, the coolness is
generated by removing the moisture from the process air using desiccant materials,
while the desiccant materials then need to be regenerated using low-grade heat, for
which solar thermal energy is one of the most promising sources.

Figure 4 illustrates the schematic of a solar-assisted desiccant cooling system
with integrated paraffin-based PCM TES. It consists of the same solar heat
collection and storage subsystem as the heating system introduced in Section 3 and
a desiccant cooling subsystem including a solid desiccant wheel, a heat recovery
ventilator, a water to air heat exchanger, an indirect evaporative cooler, an auxiliary
electric heater, two fans and the corresponding ducting system. In this system,

the solar heat collected by the evacuated tube solar collectors and/or stored in the paraffin-based PCM TES unit was used to heat the ambient air for the regeneration of the desiccant wheel, through the water to air heat exchanger. The PCM TES can also decouple the solar heat collection circuit and supply circuit, so that the retrieval of the stored thermal energy can occur by counterflow through PCM TES units during the daytime as well, if the hot water demand was higher than the hot water generated from the solar collectors. If the heat carried by the water was not sufficient for air heating, the auxiliary electric heater would be used. The desiccant wheel, together with the indirect evaporative cooler, and the heat recovery unit were used to cool the process air. In the indirect evaporative cooler, a fraction of process air was used as the secondary airflow and finally exhausted to the ambient. An ambient airflow was introduced and mixed with the return air after recovering the coolness from exhausted process air to compensate the airflow mismatch. The indoor cooling demand was satisfied by varying the airflow rate through changing the operating speed of the fans in the desiccant cooling subsystem. It is worthwhile to mention that a minimal supply airflow rate was assigned to the system operation to avoid the saturation of regeneration air after passing the desiccant wheel, and the relative humidity of the air can be further adjusted by a direct evaporative cooler before supplied to the indoor environment for space cooling.

A modelling system for this system was established using TRNSYS, in which the components for the solar heat collection and storage subsystem used were the same models as that in the heating system in Section 3. The heat exchanger, heat recovery ventilator, desiccant wheel, indirect evaporative cooler, auxiliary electric heater and fans were modelled using Type 5, Type 760, Type 716, Type 757, Type 6 and Type 111, respectively. The same typical Australian house was used to generate the building cooling load under Sydney summer weather conditions. **Table 2** also summarised the key parameters used in the numerical system performance evaluation of this system.

Figure 5 presents the performance of this solar-assisted desiccant cooling system with integrated paraffin-based PCM TES over 3 summer days. It can be seen from **Figure 5a** that the power consumption of the proposed system was from the

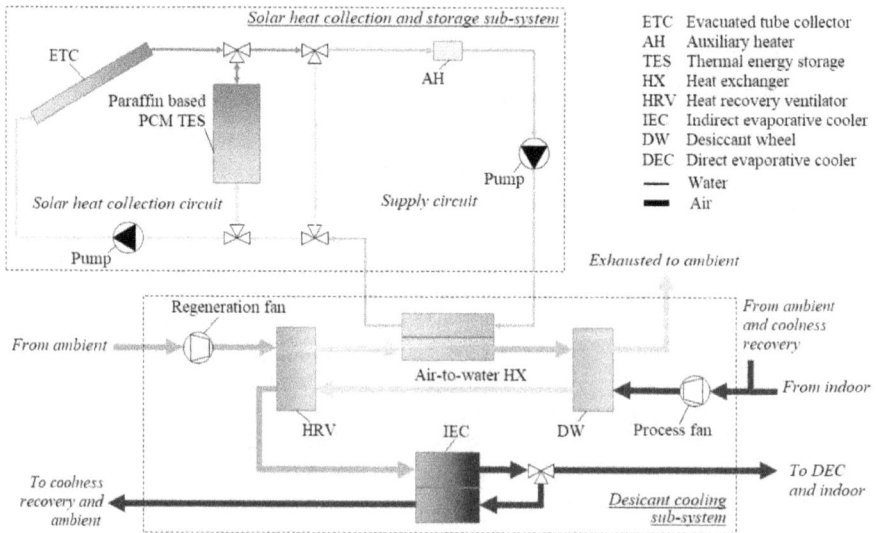

Figure 4.
Schematic of the solar-assisted radiant heating system with integrated paraffin-based PCM TES.

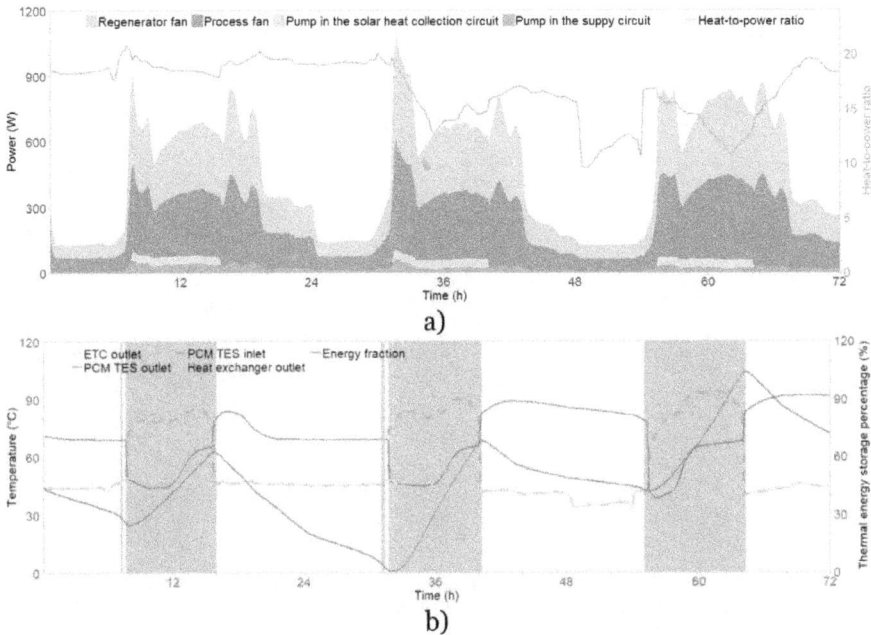

Figure 5.
Modelling results for the solar-assisted desiccant cooling system with integrated paraffin-based PCM TES.
(a) Power consumption and heat-to-power ratio. (b) Inlet and outlet water temperatures of the PCM TES.

operation of the pumps and fans, and no additional heat from the auxiliary heater was needed. The supply fan and process fan in the desiccant cooling subsystem consumed much more power (30.55 kWh) than that of the pumps (2.43 kWh) in the solar heat collection and storage subsystem. Even the fans were the major power consumers, the power consumption was much lower than the heat demand for the desiccant wheel regeneration, resulting in a high heat-to-power ratio reaching an average value of 16.55; and the corresponding average system COP reached 14.37. From **Figure 5b**, an effective charging process can be found during the PCM charging period (highlighted with the red background), while during the PCM discharging period, an outlet water temperature above 68.88°C can be achieved due to the effective thermal energy retrieval. The corresponding thermal energy storage fraction in the paraffin-based PCM fluctuated from 0.52 to 103.85% over the 3 summer test days, indicating the full utilisation of the PCM thermal energy storage capacitance.

5. Conclusions

Paraffins, as one of the main categories of phase change materials, offer the favourable phase change temperatures for solar thermal energy storage. The application of paraffin-based PCM TES in buildings can effectively rationalise the utilisation of solar energy to overcome its intermittency. Two case studies, a solar-assisted radiant heating system and a solar-assisted desiccant cooling system with integrated paraffin-based PCM TES, were presented in this chapter. The results showed that both indoor space heating and cooling can benefit from the solar TES using paraffin-based PCMs. With the assistance of the solar thermal energy storage using the paraffin-based PCMs, the energy efficiency and the heating, ventilation and air conditioning systems can be significantly improved.

Author details

Wenye Lin*, Zhenjun Ma, Haoshan Ren, Jingjing Liu and Kehua Li
Sustainable Buildings Research Centre (SBRC), University of Wollongong (UOW),
Australia

*Address all correspondence to: wenye@uow.edu.au

IntechOpen

References

[1] Bulter D. Architecture: Architects of low-energy future. Nature. 2008;**452**(7187):520-523

[2] U.S. Energy Information Administration. Available at http://www.eia.gov/consumption/ [Accessed: 26 January 2019]

[3] Perez-Lombard L, Ortiz J, Pout C. A review on buildings energy consumption information. Energy and Buildings. 2008;**40**:394-398

[4] Diana UV, Luisa Fc, Susana S, Camila B, Ksenia P. Heating and cooling energy trends and drivers in buildings. Renewable and Sustainable Energy Reviews. 2015;**41**:85-98

[5] Labat M, Virgone J, David D, Kuzanik F. Experimental assessment of a PCM to air heat exchanger storage system for building ventilation application. Applied Thermal Engineering. 2014;**66**:375-382

[6] Ma Z, Lin W, Sohel M. Nano-enhanced phase change materials for improved building performance. Renewable and Sustainable Energy Reviews. 2016;**58**:1256-1268

[7] Osterman E, Tyagi VV, Butala V, Rahim NA, Stritih U. Review of PCM based cooling technologies for buildings. Energy and Buildings. 2012;**49**:37-49

[8] Agyenim F, Hewitt N, Eames P, Smyth M. A review of materials, heat transfer and phase change problem formulation for latent heat thermal energy storage systems (LHTESS). Renewable and Sustainable Energy Reviews. 2010;**14**:615-628

[9] Soares N, Costa JJ, Caspar AR, Santos P. Review of passive PCM latent heat thermal energy storage systems towards building's energy efficiency. Energy and Buildings. 2013;**59**:82-103

[10] Beatens R, Jelle BP, Gustavsen A. Phase change materials for building applications: A state-of-the-art review. Energy and Buildings. 2010;**42**(9):1361-13689

[11] Kenisarin M, Mahkamov K. Solar energy storage using phase change materials. Renewable and Sustainable Energy Reviews. 2007;**11**:1913-1965

[12] Mehling H, Cabeza LF. Heat and Cold Storage with PCM: An up to Date Introduction into Basics and Applications. Berlin Heidelberg: Springer-Verlag; 2008

[13] Tyagi VV, Buddhi D. PCM thermal storage in buildings: a state of art. Renewable and Sustainable Energy Reviews. 2007;**11**(6):1146-1166

[14] Hasnain SM. Review on sustainable thermal energy storage technologies, part I: heat storage materials and techniques. Energy Conversion and Management. 1998;**39**(11):1127-1138

[15] Farid MM, Khudhair AM, Razack SAK, Al-Hallaj S. A review on phase change energy storage: materials and applications. Energy Conversion and Management. 2004;**45**:1597-1615

[16] Sharma A, Tyagi VV, Chen CR, Buddhi D. Review on thermal energy storage with phase change material and applications. Renewable and Sustainable Energy Reviews. 2009;**13**:318-345

[17] Kalogirou SA. Solar thermal collectors and applications. Progress in Energy and Combustion Science. 2004;**30**:231-295

[18] Tey J, Rosell JI, Ibanez M, Fernandez R. 'Solar Collector with Integrated Storage and Transparent Insulation Cover', poster at Eurosun Conference. 2002

[19] Haillot D, Nepveu F, Goetz V, Py X, Benabdelkarim M. High performance storage composite for the enhancement of solar domestic hot water systems— Part 2: Numerical system analysis. Solar Energy. 2012;**86**:64-77

[20] Al-Kayiem HH, Lin SW. Performance evaluation of a solar water heater integrated with a PCM nanocomposite TES at various inclinations. Solar Energy. 2014;**109**:82-92

[21] Haillot D, Nepveu F, Goetz V, Py X, Benabdelkarim M. High performance storage composite for the enhancement of solar domestic hot water systems— Part 1: Storage material investigation. Solar Energy. 2012;**85**:1021-1027

[22] Naghavi MS, Ong KS, Badruddin IA, Mehrali M, Silakhori M, Metselaar HSC. Theoretical model of an evacuated tube heat pipe solar collector integrated with phase change material. Energy. 2015;**91**:911-924

[23] Papadimitratos A, Sobhansarbandi S, Pozdin V, Zakhidov A, Hassanipour F. Evacuated tube solar collectors integrated with phase change materials. Solar Energy. 2016;**129**:10-19

[24] Abokersh MH, El-Morsi M, Sharaf O, Abdelrahman W. An experimental evaluation of direct flow evacuated tube solar collector integrated with phase change material. Energy. 2017;**139**:1111-1125

[25] Abokersh MH, El-Morsi M, Sharaf O, Abdelrahman. On-demand operation a compact solar water heater based on U-pipe evacuated tube solar collector combined with phase change material. Solar Energy. 2017;**155**:1130-1147

[26] Murali G, Mayilsamy K. Effect of Latent Thermal Energy storage and inlet locations on enhancement of stratification in a solar water heater under discharging mode. Applied Thermal Engineering. 2016;**106**:354-360

[27] Kousksou T, Bruel P, Cherreau G, Leoussoff V, Rhafiki TE. PCM storage for solar DHW: From an unfulfilled promise to a real benefit. Solar Energy. 2011;**85**:2033-2040

[28] Nallusamy N, Sampath S, Velraj R. Experimental investigation on a combined sensible and latent heat storage system integrated with constant/varying (solar) heat sources. Renewable Energy. 2007;**32**:1206-1227

[29] Ledesma JT, Lapka P, Domanski R, Casares FS. Numerical simulation of the solar thermal energy storage system for domestic hot water supply located in south Spain. Numerical Simulation of the Solar Thermal Energy Storage. 2013;**17**:431-442

[30] Padmaraju SAV, Viginesh M, Nallusamy N. Comparative study of sensible and latent heat storage systems integrated with solar water heating unit. Renewable Energies and Power Quality Journal. 2008;**1**:1-6

[31] Kanimozhi B, Bapu BRR. Experimental study of thermal energy storage in solar system using PCM. Advanced Materials Research. 2012;**433-440**:1027-1032

[32] Aldoss TK, Rahman MM. Comparison between the single-PCM and multi-PCM thermal energy storage design. Energy Conversion and Management. 2014;**83**:79-87

[33] Mahfuz MH, Anisur MR, Kibria MA, Saidur R, Metselaar IHSC. Performance investigation of thermal energy storage system with Phase Change Material (PCM) for solar water heating application. International Communications in Heat and Mass Transfer. 2014;**57**:132-139

[34] Karthikeyan S, Velraj R. Numerical investigation of packed bed storage unit filled with PCM encapsulated spherical containers: A comparison

between various mathematical models. International Journal of Thermal Sciences. 2012;**60**:153-160

[35] Ren HS, Ma ZJ, Lin WY, Fan WK, Li WH. Integrating photovoltaic thermal collectors and thermal energy storage systems using phase change materials with rotary desiccant cooling systems. Sustainable Cities and Society. 2018;**36**:131-143

[36] Ren HS, Ma ZJ, Lin WY, Wang SG, Li WH. Optimal design and size of a desiccant cooling system with onsite energy generation and thermal storage using a multilayer perceptron neural network and a genetic algorithm. Energy Conversion and Management. 2019;**180**:598-608

[37] Al-Abidi AA, Mat S, Sopian K, Sulaiman MY, Mohammad AT. Numerical study of PCM solidification in a triplex tube heat exchanger with internal and external fins. International Journal of Heat and Mass Transfer. 2013;**61**:684-695

[38] Al-Abidi AA, Mat S, Sopian K, Sulaiman MY, Mohammad AT. Experimental study of PCM melting in triplex tube thermal energy storage for liquid desiccant air conditioning system. Energy and Buildings. 2013;**60**:270-279

[39] Mat S, Al-Abidi AA, Sopian K, Sulaiman MY, Moohammad AT. Enhance heat transfer for PCM melting in triplex tube with internal-external fins. Energy Conversion and Management. 2013;**74**:223-236

[40] Agyenim F. The use of enhanced heat transfer phase change materials (PCM) to improve the coefficient of performance (COP) of solar powered LiBr/H2O absorption cooling systems. Renewable Energy. 2016;**87**:229-239

[41] Zhang YP, Lin KP, Yang B, Di HF, Jiang Y. Preparation, thermal

performance and application of shape-stabilized PCM in energy efficient buildings. Energy and Buildings. 2006;**38**(10):1262-1269

[42] Beckman WA. TRNSYS Reference Manual. Vol. 2. Madison: Solar Energy Laboratory; 2001

[43] Wong JPC. Development of representative dwelling designs for technical and policy purposes. Sustainable Building Innovation Laboratory: Royal Melbourne Institute of Technology University; 2013. version 6

[44] Australian Government Department of Industry. 'Representative Dwelling Models: Industry Consultation Summary Paper for Survey Participants'. 2013

[45] National House Energy Rating Scheme, Australia. Available at: www.nathers.gov.au [Accessed: 15 January 2014]

[46] Bourne S, Novoselac A. Compact PCM-based thermal stores for shifting peak cooling loads. Building Simulation. 2015;**8**:673-688

[47] Rubitherm Technology GmBH. Available at: www.rubitherm.eu [Accessed at: 15 January 2015]

[48] Jani DB, Mishra M, Sahoo PK. Solid desiccant air conditioning—A state of the art review. Renewable and Sustainable Energy Reviews. 2016;**60**:1451-1469

[49] La D, Dai YJ, Li Y, Wang RZ, Ge TS. Technical development of rotary desiccant dehumidification and air conditioning: A review. Renewable and Sustainable Energy Reviews. 2010;**14**:130-147

Chapter 5

Paraffin as Phase Change Material

Amir Reza Vakhshouri

Abstract

Nowadays, numerous problems, including the environmental problem caused by fossil fuels, have led to greater attention to the optimal use of energy and the development of renewable energy. One of the most important parts of using energy efficiently is storing it. Among the many ways introduced for energy storage, thermal energy storage, including latent heat, is among the most interesting. This storage is done with materials called phase change materials (PCMs). These materials store the energy in the form of latent heat at constant temperature during the phase transition, discussed in this chapter, and release the same stored energy in the crystallization process. These materials are mainly classified into three categories: organic, inorganic, and eutectics. Today, these materials are widely used with different properties in a variety of fields. Paraffin is one of the most important organic PCMs due to its numerous advantages that will be discussed in the following sections. From the methods of using paraffinic PCMs, two main methods, encapsulation and shape-stable PCMs, are discussed in detail. On the whole, this chapter of the book attempts to briefly discuss paraffins and their unique role in thermal energy storage systems as phase change materials.

Keywords: phase change materials, paraffin, encapsulations, shape-stable PCMs, thermal conductivity

1. Introduction

There may not be a precise background to the first discovery and application of phase change materials (PCMs). Perhaps, from the earliest days where human has acquired the intellect, he has realized the existence of these substances or, maybe, has used them without recognizing their nature. Throughout science and technology evolution, more precisely, since the heat capacity of materials and sensible or latent heats have been known, their ability to store and release thermal energy has also been considered. However, A. T. Waterman submitted the first report of discovery in the early 1900s. In recent years, scientists have paid particular attention to these materials, and their commercialization began from those years.

Perhaps the main reason for this attention was the problems caused by energy mismanagement and improper use of it. Today, inadequate energy management, especially fossil fuels, has caused many environmental and economic problems. Therefore, the necessity of efficient energy demand as well as development of renewable energies and energy storage systems is highly significant. One of the important topics in this field is the design of special energy storage equipment to other types. Energy storage not only reduces the discrepancy between energy supply and demand but also indirectly improves the performance of energy generation systems as well as plays a vital role in saving of energy by converting it into other reliable forms. Hence, this matter saves high-quality fuels and reduces energy wastes [1–3].

2. Phase change materials: an overview

Energy storage is one of the important parts of renewable energies. Energy can be stored in several ways such as mechanical (e.g., compressed air, flywheel, etc.), electrical (e.g., double-layer capacitors), electrochemical (e.g., batteries), chemical (e.g., fuels), and thermal energy storages [4].

Among several methods of energy storage, thermal energy storage (TES) is very crucial due to its advantages. TES is accomplished by changing the internal energy of materials, such as sensible heat, chemical heat, latent heat, or a combination of them.

In sensible heat storage (SHS) systems, heat can be stored by increasing the temperature of a material. Hence, this system exploits both the temperature changes and the heat capacity of the material to store energy. The amount of heat stored in this system depends on the specific heat, temperature differences, and amount of material; thus it requires a large amount of materials, whereas Latent heat storage (LHS) is generally based on the amount of heat absorbed or released during the phase transformation of a material. Lastly, In the chemical heat storage (CHS), heat is stored by enthalpy change of a chemical reaction.

Among the aforementioned heat storage systems, the LHS is particularly noteworthy. One of the special reasons is its ability to store large amount of energy at an isothermal process [5–7].

2.1 Phase change materials as thermal energy storage

Any high-performance LHS system should contain at least one of the following terms:

- Appropriate PCM with optimum melting temperature range

- Desirable and sufficient surface area proportional to the amount of heat exchange

- Optimal capacity compatible with PCM

Phase change materials perform energy storage in LHS method. In this case, a material during the phase change absorbs thermal energy from surrounding to change its state, and in the reverse process, the stored energy is released to the surrounding. PCMs initially behave likewise to other conventional materials as the temperature increases, but energy is absorbed when the material receives heat at higher temperatures and close to the phase transformation. Unlike conventional materials, in PCMs absorption or release of thermal energy is performed at a constant temperature. A PCM normally absorbs and releases thermal energy 5–14 times more than other storage materials such as water or rock [8, 9].

PCMs can store thermal energy in one of the following phase transformation methods: solid-solid, solid-liquid, solid-gas, and liquid-gas. In the solid-solid phase change, a certain solid material absorbs heat by changing a crystalline, semicrystalline, or amorphous structure to another solid structure and vice versa [10]. This type of phase change, usually called phase transitions, generally has less latent heat and smaller volume change comparing to the other types. Recently, this type of PCM has been used in nonvolatile memories [11].

Solid-liquid phase change is a common type of commercial PCMs. This type of PCM absorbs thermal energy to change its crystalline molecular arrangement to a disordered one when the temperature reaches the melting point. Unlike solid-solid, solid-liquid PCMs contain higher latent heat and sensible volumetric change. Solid-gas and liquid-gas phase changes contain higher latent heat, but their phase changes are associated with large volumetric changes, which cause many problems in TES systems [8]. Although the latent heat of solid-liquid is less than liquid-gas, their volumetric change is much lower (about 10% or less). Therefore, employing PCMs based on solid-liquid phase change in TES systems would be more economically feasible.

The overall classification of energy storage systems as well as phase change materials is given in **Figure 1**.

2.2 Classification of phase change materials

As mentioned in the previous section, despite the high thermal energy absorption capacity, PCMs in liquid-gas and solid-gas transitions have extremely high volume changes. On the other hand, solid-solid PCMs also have a lower thermal energy storage capacity. Therefore, the abovementioned PCMs, with the exception of specific cases, have not received much attention to commercialization. Currently, the most common type of transition that has been mass-marketed is solid-liquid PCMs. The classification of phase change materials is schematically given in **Figure 1**. Solid-liquid PCMs are generally classified as three general organics, inorganic, and eutectics [12, 13]. However, in some references they are classified into two major organics and inorganics.

2.2.1 Inorganic PCMs

Inorganic PCMs mainly have high capacity for thermal energy storage (about twice as much as organic PCMs) as well as have higher thermal conductivity. They are often classified as salt hydrates and metals.

Salt hydrates are the most important group of inorganic PCMs, which is widely employed for the latent heat energy storage systems. Salt hydrates are described as a mixture of inorganic salts and water (AB × nH_2O). The phase change in salt hydrates actually involves the loss of all or plenty of their water, which is roughly equivalent to the thermodynamic process of melting in other materials.

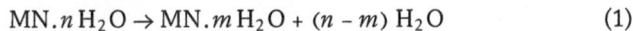

$$MN.n\,H_2O \rightarrow MN.m\,H_2O + (n-m)\,H_2O \qquad (1)$$

$$MN.n\,H_2O \rightarrow MN + n\,H_2O \qquad (2)$$

At the phase transition, the hydrate crystals are subdivided into anhydrous (or less aqueous) salt and water. Although salt hydrates have several advantages, some deficiencies make restrictions in their application. One of these problems is incongruent melting behavior of salt hydrates. In this problem the released water from dehydration process is not sufficient for the complete dissolution of the salts. In this case, the salts precipitate and as a result phase separation occurs. In order to prevent this problem, an additional material such as thickener agent is added to salt hydrates. Another major problem with salt hydrates is the supercooling phenomenon. In this phenomenon, when crystallization process occurs, the nucleus

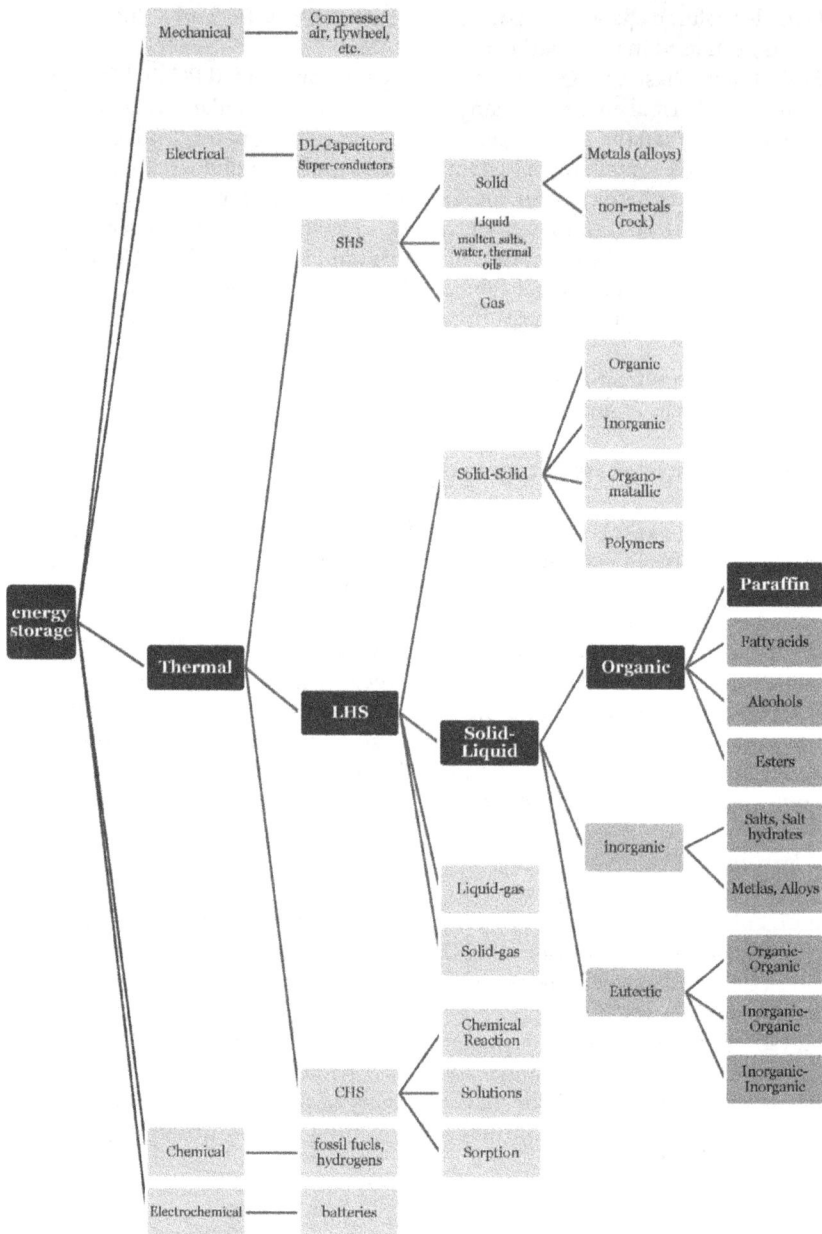

Figure 1.
Overview of energy storage and classification of phase change materials.

formation is delayed; therefore, even at temperatures below freezing, the material remains liquid [7, 11, 14].

Overall, the most attractive properties of salt hydrate are (i) high alloy latent temperature, (ii) relatively high thermal conductivity (almost two to five times more than paraffin), and (iii) small volume changes in melting. They are also very low emitting and toxic, adaptable to plastic packaging, and cheap enough to use [15].

Metals are another part of the inorganic PCMs. Perhaps the most prominent advantages of metals are their high thermal conductivity and high mechanical properties. Metals are available over a wide range of melting temperatures. They are also used as high-temperature PCMs.

Some metals such as indium, cesium, gallium, etc. are used for low-temperature PCMs, while others such as Zn, Mg, Al, etc. are used for high temperatures. Some metal alloys with high melting points (in the range of 400–1000°C) have been used for extremely high temperature systems. These metal alloys as high-temperature PCMs can be used in the field of solar power systems [16, 17]. They can also be used in industries that require temperature regulation in furnaces or reactors with high operating temperatures.

2.2.2 Organic PCMs

Perhaps the most important fragment is the organic PCMs. Organic PCMs show no change in performance or structure (e.g., phase separation) over numerous phase change cycles. In addition, supercooling phenomena cannot be observed in organic PCMs. The classification of organic PCMs is unique. This division is mainly based on their application contexts. In general, they are classified into two major paraffin and non-paraffin sections.

Paraffins are the most common PCMs. Since this book is about paraffin, to avoid duplication, this section will briefly discuss the chemistry (structure and properties) of paraffin, but their ability as phase change materials will be reviewed in detail.

Non-paraffinic organic PCMs are known to be the most widely used families. In addition to their different properties compared to paraffins, they have very similar properties to each other. Researchers have used various types of ether, fatty acid, alcohol, and glycol as thermal energy storage materials. These materials are generally flammable and less resistant to oxidation [18–20].

Although non-paraffin organic PCMs have high latent heat capacity, they have weaknesses such as flammability, low thermal conductivity, low combustion temperatures, and transient toxicity. The most important non-paraffinic PCMs are fatty acids, glycols, polyalcohols, and sugar alcohols.

Fatty acids [$CH_3(CH_2)_{2n}COOH$] also have high latent heat. They can be used in combination with paraffin. Fatty acids exhibit high stability to deformation and phase separations for many cycles and also crystallize without supercooling. Their main disadvantages are their costs. They are 2–2.5 times more expensive than technical grade paraffins. Unlike paraffins, fatty acids are of animal or plant origin. Their properties are similar to those of paraffins, but the melting process is slower. On the other hand, they are moderately corrosive as well as generally odorous [21].

2.2.3 Eutectics

A eutectic contains at least two types of phase change materials. Eutectics have exceptional properties. In eutectics, the melting-solidification temperatures are generally lower than the constituents and do not separate into the components through the phase change. Therefore, phase separation and supercooling phenomena are not observed in these materials.

Eutectics typically have a high thermal cycle than salt hydrates. Inorganic-inorganic eutectics are the most common type of them. However, in recent studies, organic-inorganic and organic-organic varieties have received more attention. The major problem of eutectics is their commercialization. Their cost is usually two to three times higher than commercial PCMs [22, 23].

Some of the above PCMs and their thermal properties, which are competitive with paraffins in terms of latent heat capacity, are summarized in **Table 1**.

Type of PCMs	Materials	Melting point (°C)	Latent heat (kJ/kg)	Density* (kg/m³)	Thermal conductivity (W/mK)**	Ref.
Inorganic salt hydrates	$LiClO_3 \cdot 3H_2O$	8	253	1720		[24, 25]
	$K_2HPO_4 \cdot 6H_2O$	14	109			[24]
	$Mn(NO_3)_2 \cdot 6H_2O$	25.8	126	1600		[14, 25]
	$CaCl_2 \cdot 6H_2O$	29.8	191	1802	1.08	[24, 25]
	$Na_2CO_3 \cdot 10H_2O$	32-34	246-267			[14, 24]
	$Na_2SO_4 \cdot 10H_2O$	32.4	248, 254	1490	0.544	[14, 26]
	$Na_2HPO_4 \cdot 12H_2O$	34-35	280	1522	0.514	[15, 26]
	$FeCl_3 \cdot 6H_2O$	36-37	200, 226	1820		[25, 26]
	$Na_2S_2O_3 \cdot 5H_2O$	48-49	200, 220	1600	1.46	[15, 26]
	$CH_3COONa \cdot 3H_2O$	58	226, 265	1450	1.97	[15, 26]
Non-paraffinic organic PCMs	Fatty acids					
	Formic acid	8.3	247	1220	—	[1, 25]
	n-Octanoic acid	16	149	910	0.148	[21, 27]
	Lauric acid	43.6	184.4	867		[21, 25]
	Palmitic acid	61.3	198	989	0.162	[21, 27]
	Stearic acid	66.8	259	965	0.172	[21, 25]
	Polyalcohols					
	Glycerin	18	199	1250	0.285	[1, 25]
	PEG E600	22	127.2	1126	0.189	[27]
	PEG E6000	66	190	1212		[27]
	Xylitol	95	236	1520	0.40	[28]
	Erythritol	119	338	1361	0.38	[28]
	Others					
	2-Pentadecanone	39	241			[1, 25]
	4-Heptadekanon	41	197			[1, 25]
	D-Lactic acid	52-54	126, 185	1220		[1, 25]

Type of PCMs	Materials		Melting point (°C)	Latent heat (kJ/kg)	Density* (kg/m³)	Thermal conductivity** (W/mK)	Ref.
Eutectics	O-O, O-I, I-I***	$CaCl_2 \cdot 6H_2O + MgCl_2 \cdot 6H_2O$	25	127	1590		[27]
		$Mg(NO_3)_2 \cdot 6H_2O + MgCl_2 \cdot 6H_2O$	59	144	1630	0.51	[27]
		Trimethylolethane + urea	29.8	218			[21]
		$CH_3COONa \cdot 3H_2O$ + Urea (60:40)	31	226			[27]
	Metals	Mg-Zn (72:28)	342	155	2850	67	[16, 17]
		Al-Mg-Zn (60:34:6)	450	329	2380		[16, 17]
		Al-Cu (82:18)	550	318	3170		[16, 17]
		Al-Si (87.8:12.2)	580	499	2620		[16, 17]

*At 20°C.
**Just above melting point (liquid phase).
***Inorganic-inorganic (I-I), organic-inorganic (O-I), and organic-organic (O-O).

Table 1.
Thermophysical properties of some common PCMs with high latent heat.

3. Paraffin-based phase change materials

Paraffin is usually a mixture of straight-chain n-alkanes with the general formula CH_3-$(CH_2)_n$-CH_3. However, in some cases, paraffin is used as another name for alkanes. Gulfam R. et al. in their article have classified paraffins based on the number of carbon atoms as well as their physical states. According to this classification, at room temperature, 1–4 numbers of carbons refer to pure alkanes in a gas

Materials	Melting point (°C)	Latent heat (kJ/kg)	Density* (kg/m³)	Thermal conductivity** (W/mK)
n-Tetradecane (C$_{14}$)	6	228–230	763	0.14
n-Pentadecane (C$_{15}$)	10	205	770	0.2
n-Hexadecane (C$_{16}$)	18	237	770	0.2
n-Heptadecane (C$_{17}$)	22	213	760	0145
n-Octadecane (C$_{18}$)	28	245	865	0.148
n-Nonadecane (C$_{19}$)	32	222	830	0.22
n-Eicosane (C$_{20}$)	37	246		
n-Henicosane (C$_{21}$)	40	200, 213	778	
n-Docosane (C$_{22}$)	44.5	249	880	0.2
n-Tricosane (C$_{23}$)	47.5	232		
n-Tetracosane (C$_{24}$)	52	255		
n-Pentacosane (C$_{25}$)	54	238		
n-Hexacosane (C$_{26}$)	56.5	256		
n-Heptacosane (C$_{27}$)	59	236		
n-Octacosane (C$_{28}$)	64.5	253		
n-Nonacosane (C$_{29}$)	65	240		
n-Triacontane (C$_{30}$)	66	251		
n-Hentriacontane (C$_{31}$)	67	242		
n-Dotriacontane (C$_{32}$)	69	170		
n-Triatriacontane (C$_{33}$)	71	268	880	0.2
Paraffin C$_{16}$-C$_{18}$	20–22	152		
Paraffin C$_{13}$-C$_{24}$	22–24	189	900	0.21
RT 35 HC	35	240	880	0.2
Paraffin C$_{16}$-C$_{28}$	42–44	189	910	
Paraffin C$_{20}$-C$_{33}$	48–50	189	912	
Paraffin C$_{22}$-C$_{45}$	58–60	189	920	0.2
Paraffin C$_{21}$-C$_{50}$	66–68	189	930	
RT 70 HC	69–71	260	880	0.2
Paraffin natural wax 811	82–86	85		0.72 (solid)
Paraffin natural wax 106	101–108	80		0.65 (solid)

*At 20°C.
**Just above melting point (liquid phase).

Table 2.
Thermophysical properties of n-paraffins and commercial paraffinic PCMs [1, 24, 25].

phase, 5–17 carbons are liquid paraffins, and more than 17 is known as solid waxes. These waxy solids refer to a mixture of saturated hydrocarbons such as linear, iso, high branched, and cycloalkanes [29]. Generally, paraffin-based PCMs are known as waxy solid paraffins. Commercial paraffins contain mixture of isomers, and therefore, they have a range of melting temperatures.

Paraffins typically have high latent heat capacity. If the length of the chain increases, the melting ranges of waxes also increase, while the latent heat capacity of melting is not subject to any particular order (**Table 2**).

In general, paraffin waxes are safe, reliable, inexpensive, and non-irritating substances, relatively obtained in a wide range of temperatures. As far as economic issues are concerned, most technical grade waxes can be used as PCMs in latent heat storage systems. From the chemical point of view, paraffin waxes are inactive and stable. They exhibit moderate volume changes (10–20%) during melting but have low vapor pressure.

The paraffin-based PCMs usually have high stability for very long crystalliza-tion-melting cycles. **Table 2** illustrates the thermal properties of some paraffin waxes.

Besides the favorable properties, paraffins also show some undesirable proper-ties such as low thermal conductivity, low melting temperatures, and moderate-high flammability. Some of these disadvantages especially thermal conductivity and flammability can be partially eliminated with the help of additives or paraffin composites.

4. Methods for using paraffin-based PCMs (PPCMs)

Measures must be taken to make the solid-liquid PCMs usable. For this purpose, there are several methods for stabilizing the shapes of paraffinic PCMs. Two main methods of them are discussed below.

4.1 Encapsulation of PPCMs

Encapsulation is generally a worthy method to protect and prevent leakage of PCMs in the liquid state. The capsules consist of two parts, the shell and the core. The core part contains PCMs, whereas the shell part is usually composed of poly-meric materials with improved mechanical and thermal properties. The shell part plays the role of protection, heat transfer, and sometimes preventing the release of toxic materials into the environment. In these cases, the shell must have appropriate thermal conductivity. Polymeric shells are also commonly used in encapsulating PPCMs. The choice of core part depends on its application field. The encapsulation of PPCMs is classified into three major parts: bulk or macroencapsulation, microen-capsulation, and nano-encapsulation.

Macroencapsulation is one of the simplest ways to encapsulate paraffins. This method has a lower cost than other methods. These products are used in transpor-tation, buildings, solar energy storage systems, and heat exchangers. Sometimes metals are also used as shell materials [30].

In order to increase the efficiency of heat transfer in these types of capsules, either the size of the capsules should be appropriately selected or suitable modifiers should be used. In general, the smaller the diameter of spherical capsules or cylinders, the bet-ter the heat transfer. In some cases, metal foams are used to improve the heat transfer properties of paraffin. Aluminum and copper open-cell foams are among the most studied, whereas, in other cases metal oxides, metals and graphite are used [30, 31].

There are various forms of macroencapsulation, such as ball shape, spherical shape, cylindrical, flat sheets, tubular, etc. [31]. Cylindrical tubes are one of the famous forms of macroencapsulated PPCMs. This type of encapsulation is most commonly used in buildings or in solar energy storage systems.

Most of the research carried out on macroencapsulated PPCMs has been focused on improving their thermal conductivity. In one of these studies, different metal oxide nanoparticles such as aluminum oxide, titanium oxide, silicon oxide, and zinc oxide were used to improve the thermal conductivity of paraffin. The results show that titanium oxide performs better under the same conditions than the other oxides [32]. In a similar study, copper oxide nanoparticles were used to improve thermal conductivity and performance of paraffin in solar energy storage systems [33]. In some studies, graphite flakes and expanded graphite have also been used as improving agent for heat conductivity [31].

Hong et al. have used polyethylene terephthalate pipes as a shell for paraffin. In this macroencapsulated system, introduced as cylinder modules, float stone has been added to paraffin as an enhancer of thermal conductivity. In this study, the effect of various parameters such as pipe diameter on heat transfer is investigated, and the results of experimental section are compared with modeling [34].

D. Etansova et al. studied numerical computation and heat transfer modeling of paraffin-embedded stainless steel macroencapsulates for use in solar energy storage systems. In this study, the effect of geometric size and shape on heat transfer was investigated [35].

Microencapsulation of PCMs is another suitable way to improve efficiency and increase thermal conductivity. The size of the microencapsulates usually ranges from 1 μm to 1 mm. Microencapsulation of paraffins is a relatively difficult process, but it performs better than macroencapsulates. This is due to increased contact surface area, shorter discharge and loading times, and improved thermal conductivity. Different materials are used for the shell part of the microencapsulates.

In general, there are two major physical and chemical methods for microencapsulation. The most important physical methods are fluidized bed, spray dryer, centrifuge extruder, and similar processes. However, chemical methods are often based on polymerization. The most important techniques include in situ suspension and emulsion polymerization, interfacial condensation polymerization, and sol-gel method. The latter is sometimes known as the physicochemical method [12, 29].

In the suspension or emulsion polymerization method, the insoluble paraffin is first emulsified or suspended in a polar medium, which is predominantly aqueous phase, by means of high-speed stirring. Surfactants are used to stabilize the particles. Then, lipophilic monomers are added to the medium, and the conditions are prepared for polymerization. This polymer, which is insoluble in both aqueous and paraffin phases, is formed on the outer surface of paraffin particles and finally, after polymerization, encapsulates the paraffin as a shell. The size of these capsules depends on the size of emulsion or suspension of paraffin droplets. Sometimes certain additives are added to the medium to improve some of the polymer properties. For instance, in some studies, polyvinyl alcohol (PVA) has been added to the medium with methyl-methacrylate monomer, which is known as one of the most important shell materials. As a result, paraffin has been encapsulated by PVA modified polymethyl methacrylate (PMMA). Adding this modifier forms a smooth surface of the microencapsulates [36, 37].

In the interfacial method, soluble monomers in the organic phase with other monomers in the aqueous phase at the droplet interface form a polymer that precipitates on the outer layer of the organic phase.

The sol-gel method is a multi-step procedure. In this method, firstly, an organosilicon compound such as tetraethoxysilane (TEOS) is hydrolyzed in an acidic medium at low pH. The prepared homogenous solution is known as the sol part. Then, the paraffin emulsion is prepared in an aqueous medium and stabilized by special emulsifiers. Actually, these emulsifiers are the first layer of the shell. Subsequently, the sol solution is slowly added to the aqueous phase containing paraffin. The silicon compounds containing OH groups (silanols) form hydrogen bonding with polar side of emulsifiers, and finally the condensation process is carried out on the first layer interface. As a result, paraffin microencapsulates with an inorganic material that is often silica. Silica is one of the significant materials used as a shell for micro and nano-encapsulation. Silica has high thermal conductivity and on the other hand has better mechanical properties than some polymers [38–41].

As mentioned, most of the materials used to microencapsulation are polymers. The main polymers used as shell materials are polymethyl methacrylate [42], polystyrene [43], urea-formaldehyde [44], urea-melamine-formaldehyde [45], polyaniline [46], etc. However, in many cases, these polymers are used in modified form. For example, polymethyl methacrylate modified with polyvinyl alcohol or with other methacrylates [36, 37], polystyrene copolymers [47], and melamine modified-formaldehyde with methanol [48] can be considered. **Table 3** shows the most common polymers used as shell materials.

In addition to the aforementioned microencapsulation approaches, which mainly form polymeric materials as shells, other materials have been also recommended. For example, Singh and colleagues have used silver metal as a shell for paraffin microencapsulates. They first emulsified paraffin into small particles in water and then converted silver salts to metallic silver via an in situ reduction reaction. The average particle size of 329 μm has been reported, and the thermal properties of paraffin have been investigated using DSC and TGA. This type of metal shell microencapsulates has been suggested for use in microelectronics heat management systems [49].

There are several techniques to study the properties of micro and nano-encapsulates. In all studies, differential scanning calorimetry (DSC) and thermogravimetric analysis (TGA) have been used to determine the thermal properties of PPCMs, such as enthalpy of fusion, melting temperature, weight loss, degradation, etc. Various methods such as XRD, FTIR, and ^{12}C NMR have been used to study the structure and chemical composition of PPCMs. The morphology and diameters of the microcapsules have often been studied by scanning electron microscopy (SEM) and particle size analyzer.

The latter technique is used to study the influence of different variables on the diameter of the microcapsules. One of these variables is the effect of stirring speed on emulsification of paraffin. The results of some studies show that higher stirring speed of emulsification process leads to decrease of the mean size of paraffin droplets [48].

Along with studies on the type of microcapsules, many studies have been conducted to improve thermal conductivity and mechanical properties of microencapsulates. Part of these studies has been dedicated to the effect of graphene and graphene oxide on the improvement of thermal conductivity [51]. L. Zhang et al. investigated the effect of graphene oxide on improving the mechanical properties and leakage protection as well as improving the thermal conductivity of melamine-formaldehyde as shell materials of PPCM microencapsulates [52]. In another part of studies, metals and metal oxides have been used. For example, 10 and 20 wt% of nanomagnetite (Fe_3O_4) with particle size from 40 to 75 nm increase the thermal conductivity by 48 and 60%, respectively [53]. Also, addition of TiO_2 and Al_2O_3 nanoparticles in a mass fraction of 5% with respect to PPCM at the size range of 30–60 nm increases the thermal conductivity by 40 and 65%, respectively [54].

Core material PPCM	Shell material	Encapsulation method	Particle size (μm)	Recommended application	Ref
n-Nonadecane	Polymethyl methacrylate	Emulsion	~ 8	Smart building and textiles	[42]
n-Heptadecane	Polystyrene	Emulsion	<2	General fields	[43]
Commercial paraffin wax	Polystyrene-co-PMMA	Suspension	~ 20		[50]
Commercial RT21	PMMA	Suspension	20–40		[36]
Commercial RT21	PMMA modified with PVA	Emulsion	15	Building	[37]
Commercial paraffin wax	Polyaniline	Emulsion	<1		[46]
Commercial paraffin wax	Urea-formaldehyde	In situ	~ 20		[44]
n-Octadecane, n-nonadecane	Urea-melamine-formaldehyde	In situ	0.3-0.6		[45]
Commercial paraffin wax	Methanol-melamine-formaldehyde	In situ	10–30	Building	[48]
Commercial paraffin wax	Silica	Sol-gel	4–10	Textile	[38]
Commercial paraffin wax	Silica	Sol-gel	0.2–0.5		[39]
n-Octadecane	Silica	Sol-gel	7–16		[40]
n-Pentadecane	Silica	Sol-gel	4–8		[41]

Table 3.
Common materials for microencapsulation of PPCMs.

Nano-encapsulation of PPCM is very similar to the microencapsulation process. However, these types of encapsulation specific techniques, such as ultrasonic, are used to adjust the size of the paraffin droplets to less than 1 micron. In the next step, using the chemical methods mentioned in the microencapsulation method, the shell formation is performed. The most common method for nano-encapsulation is the emulsion polymerization method. However, although limited, interfacial and sol-gel methods have also been reported.

4.2 Shape-stable PPCMs

In recent years, research on polymeric matrix-based shape-stable PCMs has gained great importance. Among these types of phase change materials, the paraffin-polymer composite is particularly attractive. The combination of paraffin and polymers as new PCMs with a unique controllable structure can be widely used. This compound remains solid at paraffin melting point and even above without any softening, which is why this type of PCM is called shape-stable. These materials are well formed and have high-energy absorption capacity; hence they can be widely used as stable PCMs with specific properties. On the other hand, some problems such as high cost and difficulty of encapsulating processes could be resolved. Despite these advantages, some common disadvantages such as low thermal stability, low thermal conductivity, and relatively high flammability can restrict their application, particularly in building materials. For this reason, further studies are required to eliminate these disadvantages and improve the properties of these materials. A large part of research is relevant to increase or improve their thermal conductivity, flame retardation, and thermophysical and mechanical properties. Suitable additives are proposed to improve these properties [55, 56].

In some articles, a simple method involves mixing-melting of polyethylene and paraffin, consequently cooling the composite, or using a simple twin extruder to prepare a shape-stable PCM has been reported [57, 58]. When this compound contains sufficient polymer, a homogeneous mixture remains solid at temperatures above the melting point of paraffin and below the polymer melting point. During the preparation of these composites, no chemical reaction or chemical bonds are formed between the polymers and paraffin; therefore these types of compounds are considered as physical mixtures. Shape-stable PPCMs can be used in all previously described areas. Due to the thermoplastic properties of these composites, it is possible to melt and crystalize them for many cycle numbers. Shape-stable PPCMs have several advantages over other PCMs. They are also nontoxic and do not require high-energy consumption during production process.

Inaba and Tu [59] developed a new type of shape-stable PPCM and determined their thermophysical properties. These materials can be used without encapsulation. Feldman et al. [60] prepared plates of shape-stable PCM and determined their high thermal energy storage capacity when used in small chambers. In this type of polymer-based plates, fatty acids are used as PCMs that absorb or releases large amounts of heat during melting and solidification, without altering the composition of the shape-stable PCM. The same researchers determined the role of polymer-PCM sheets in stabilizing the shape and size of the plates when PCM was liquefied. The composition of paraffin and high-density polyethylene (HDPE) has been studied by Lee and Choi [61] and has been introduced as a shape-stable energy storage material. In this study, the amount of energy stored by the mentioned composites is also studied. They also studied the morphology of the high-density polyethylene crystal lattice (HDPE) and its effect on paraffin through scanning electron microscopy and optical microscopy (OM) analysis. On the other hand, they also reported of high thermal energy storage capacity of the prepared paraffin/HDPE-based shape-stable PCMs. Hong and Xin-Shi [62] synthesized polyethylene-paraffin as a shape-stable PCM and characterized its morphology and structure by scanning electron microscopy and its latent heat of melting by differential scanning calorimetry. In this study, a composition consisting of 75% paraffin as a cheap, effective, easy-to-prepare, low-temperature shape-stable PPCM is recommended. In another study, Xiao et al. [63] prepared a shape-stable PCM based on the composition of paraffin with a thermoplastic elastomer (styrene butadiene rubber) and determined its thermal properties. The obtained results show that the stable mixture has the phase changing property and the amount of latent heat of melting stored in this compound is estimated to be 80% of pure paraffin. In another part of this study, the thermal conductivity of PCMs was significantly increased by using graphite.

Despite the above benefits, some disadvantages of shape-stable PPCMs are also reported. One of the major problems is the softening and paraffin leakage phenomenon at elevated temperatures. Seiler partly resolved this problem by adding a different ratio of silica and copolymers to the polyethylene-paraffin composition [64]. Another problem is the low thermal conductivity of the polyethylene-paraffin compound. A lot of research has been conducted to increase this property. A. Sari [65] prepared two types of paraffin with different melting temperatures (42–44°C and 56–58°C) and combined each with HDPE as phase modifier. By addition of 3% expanded graphite, the thermal conductivity of composites increased by 14 and 24%, respectively. Zhang et al. [66] developed new PCMS based on graphite and paraffin with high thermal energy storage capacity and high thermal conductivity. Zhang and Ding et al. [67] have used various additives such as diatomite, Wollastonite, organic modified bentonite, calcium carbonate, and graphite to improve the thermal conductivity of shape-stable PCMs.

It should be noted that metal particles and metal oxides due to their higher thermal conductivity are widely used to improve this property of PCMs. One of the materials that has received more attention in recent years is alumina. Aluminum oxide nanoparticles were added to paraffin to increase its thermal conductivity in both liquid and solid states [57, 68]. This compound coupled with its high thermal conductivity is cheaper and more abundant than other metal oxides.

Another problem with shape-stable PPCMs is their flammability. The effect of various additives has been studied by scientists to eliminate this problem. One of the most effective of these substances is halogenated compounds, but they cause environmental pollution and also release toxic compounds while burning. Researchers have used hybrid and environmentally friendly materials to enhance the durability of flame retardant materials. They studied the effect of clay nanoparticles and organo-modified montmorillonite. Adding these materials not only increases their resistance to burning but also increases their mechanical and thermal properties [69–71]. In another study, Y. Cai et al. added paraffin, HDPE, and graphite, then added ammonium polyphosphate and zinc borate separately, and studied their resistance to burning. The results show that the addition of ammonium polyphosphate decreases flammability, while zinc borate increases the flammability risk [72]. One of the most interesting and harmless fire retardant compounds is metal hydroxides, especially aluminum hydroxide, magnesium hydroxide, or their combination [73–75].

Some researchers have used other advanced materials as supporting materials to prepare shape-stable PPCMs instead of using the polymer matrix [76–78]. Rawi et al. used acid-treated multi-walled carbon nanotubes (A-CNT). They reported that adding 5% by weight A-CNT to paraffin decreases 25% of the latent heat while increasing heat conductivity up to 84% [79]. Y. Wan et al. used pinecone biochar as the supporting matrix for PCMs. They prepared shape-stable PCM materials at different ratios and studied the leakage behavior. The optimal ratio is suggested as 60% of the PCM. For the above ratio, no PCM leakage was observed after the melting temperature. The results showed that the thermal conductivity of the same ratio shape-stable PCM increased by 44% compared to the pure PCM [80].

5. Criteria for selection of PCMs and application fields of PPCMs

PCMs are available in a wide range of desired temperature ranges. Obviously, a PCM may not have all the properties required to store heat energy as an ideal material. Therefore, it would be more appropriate to use these materials in combination with either other PCMs or various additives to achieve the required features. However, as latent heat storage materials, while using PCMs, the thermodynamic, kinetic, and chemical properties as well as the economic and availability issues of them must be taken into account. Employed PCMs must have the optimum phase change temperature. On the other hand, the higher the latent heat of the material, the lower its physical size. High thermal conductivity also helps to save and release energy. From the physical and kinetic point of view, the phase stability of PCMs during melting and crystallization contributes to optimum thermal energy storage. Their high density also enables high storage at smaller material sizes. During phase change, smaller volume changes and lower vapor pressures are appropriate for continuous applications.

H. Nazir et al. in their review article [12] have explained the criteria for selection of PCMs as a pyramid. In this pyramid, at the bottom, known as the fundamentals, there are several items such as cost, regularity compliance, and safety. In the next section, the thermophysical properties such as energy storage capacity and runtime are discussed. In the upper section, reliability and operating environment consist of degradation, cycle life, shelf life, and thermal limits are reflected. Finally, at the top section of pyramid, user perception and convenience are located. These criteria help us to find a proper PCM for certain application fields.

These criteria may also be extended to paraffinic PCMs. Nowadays, paraffinic PCMs (PPCMs) are widely used as thermal energy storage materials, including solar energy storage systems, food industries, medical fields, electrical equipment protection, vehicles, buildings, automotive industries, etc. [24, 29, 81–85].

Generally, application fields of PPCMs can be considered in two main sections: thermal protection and energy storage purposes. The major difference between these two areas of application is in thermal conductivity of the PPCMs.

Protection and transportation of temperature-sensitive materials is one the mentioned area. Sometimes a certain temperature is required to transport sensitive medicines, medical equipment, food, etc. In all cases, using of PPCMs would be appropriate as they can regulate and stabilize the temperature over a given range. Similarly, in sensitive electrical equipment, these materials are also essential to prevent the maximum operating temperature. On the other hand, they can be used to prevent possible engine damage at high temperatures [86, 87].

One of the studies related to these issues is the use of paraffin containing heavy alkanes to protect electronic devices against overheating. In this study, paraffin has been used as a protective coating for the resistor chip, and its effect on cooling of the devices has been investigated. Experimental results show that paraffin coating increases the relative duration of overheating by 50 to 150% over the temperature range of 110–140°C [88]. In another study, a mixture of paraffin and polypropylene has been used as an overheating protector in solar thermal collectors [89].

However, *energy storage purposes* are the most important part of PPCM application. In general, PCMs act as passive elements and therefore do not require any additional energy source. Most studies on the application of energy storage properties of PPCMs have been confined to buildings, textiles, and solar systems. In the following, building applications will be further attended.

One of the main drawbacks of lightweight building materials is their low thermal storage capacity, which results in extensive temperature fluctuations as a result of intense heating and cooling. Therefore, PPCMs have been used in buildings due to their ability to regulate and stabilize indoor temperatures at higher or lower outdoor temperatures [90].

Generally, PPCMs in buildings are used as thermal energy storage at daytime peak temperature, and they released the stored energy at night when temperatures are low. The result of this application is to set the comfort condition for a circadian period. This application minimizes the amount of energy consumed for cooling during the day and warming up at night.

In contrast, in order to stabilize the ambient conditions at low temperatures, some special PCMs are also used in air conditioner systems. In this case, cool air is stored during the night and released into the warm hours of the day.

Y. Cui et al. [91] in a review article categorized PPCM application methods based on their location of use such as PCMs in walls, floor heating systems, ceiling boards, air-based solar heating systems, free cooling systems (with ventilation systems), and PCM shutter (in windows). Both types of encapsulation and shape-stable PPCMs could be used in all of the above classification of building applications. Sometimes these materials can be added directly to concrete, gypsum, etc. [90, 92–95].

In order to increase the performance of PPCMs in this application field, great deals of studies have also been done on improving their thermal conductivity. On the other hand, extensive research into safety issues has been done to reduce the flammability of PPCMs by adding flame retardants to these materials.

Overall, these studies cover the importance of using PPCMs in heating and cooling as well as indicate the general characteristics, advantages, and disadvantages of these materials used for thermal storage in buildings.

6. Conclusion

It is clear that at this time, where renewable energy is particularly important, the use of PPCMs is on the rise. As it has been mentioned, PPCMs have many application fields due to their advantages. For example, they can be used in the construction, pharmaceutical and medical industries, textiles, automobiles, solar power systems, transportation, thermal batteries, heat exchangers, and so on.

This chapter of the book has attempted to focus more on how to use paraffins. For this reason, two methods, namely, encapsulation and shape-constant, have been widely discussed. In addition, improving their weak properties such as thermal conductivity and flammability has also been studied. Depending on the benefits of paraffins, new applications are suggested every day. Extensive studies are underway on other new applications in recent years.

Author details

Amir Reza Vakhshouri
Department of Chemical Engineering, Baku Higher Oil School, Baku, Azerbaijan

*Address all correspondence to: amir.vakhshouri@bhos.edu.az

IntechOpen

References

[1] Sharma A, Tyagi V, et al. Review on thermal energy storage with phase change materials and applications. Renewable and Sustainable Energy Reviews. 2009;**13**:318-345

[2] Alieva RV, Vakhshouri AR, et al. Heat storage materials based on polyolefins and low molecular weight waxes. Plastics (Russian language). 2012;**10**:42-46

[3] Gunther E, Mehling H, Werner M. Melting and nucleation temperatures of three salt hydrate phase change materials under static pressures up to 800 MPa. Journal of Applied Physics. 2007;**40**:4636-4641

[4] Cabeza LF, Castell A, Barreneche C, et al. Materials used as PCM in thermal energy storage in buildings: A review. Renewable and Sustainable Energy Reviews. 2011;**15**(3):1675-1695

[5] Wang C, Zhu Y. Chapter 2: Experimental and numerical studies on phase change materials. In: Phase Change Materials and Their Applications. London, UK: IntechOpen; 2018

[6] Regin FA, Solanki SC, Saini JS. Heat transfer characteristics of thermal energy storage system using PCM capsules: A review. Renewable and Sustainable Energy Reviews. 2008;**12**(9):2438-2458

[7] Farid MM, Khudhair AM, et al. A review on phase change energy storage: Materials and applications. Energy Conversion and Management. 2004;**45**(9-10):1597-1615

[8] Li WD, Ding EY. Preparation and characterization of crosslinking PEG/MDI/PE copolymer as solid–solid phase change heat storage material. Solar Energy Materials and Solar Cells. 2007;**91**(9):764-768

[9] Assis E, Katsman L, Ziskind G, Letan R. Numerical and experimental study of melting in a spherical shell. International Journal of Heat and Mass Transfer. 2007;**50**(9-10):1790-1804

[10] Fallahi A, Guldentops G, Tao M, et al. Review on solid-solid phase change materials for thermal energy storage: Molecular structure and thermal properties. Applied Thermal Engineering. 2017;**127**:1427-1441

[11] Mhadhbi M. Introductory chapter: Phase change material. In: Phase Change Materials and Their Applications. London, UK: IntechOpen; 2018

[12] Nazir H, Batool M, et al. Recent developments in phase change materials for energy storage applications: A review. International Journal of Heat and Mass Transfer. 2019;**129**:491-523

[13] Raoux S, Welnic W, Ielmini D. Phase change materials and their application to nonvolatile memories. Chemical Reviews. 2010;**110**:240-267

[14] Xie N, Huang Z, et al. Inorganic salt hydrate for thermal energy storage. Applied Sciences. 2017;**7**:1317

[15] Sharma SD, Sagara K. Latent heat storage materials and systems: A Review. International Journal of Green Energy. 2005;**2**:1-56

[16] Khare S, Dell Amico M, et al. Selection of materials for high temperature latent heat energy storage. Solar Energy Materials & Solar Cells. 2012;**107**:20-27

[17] Nomura T, Zhu C, et al. Microencapsulation of metal-based phase change material for high-temperature thermal energy storage. Scientific Reports. 2015;**5**:9117

[18] Stritih U. Heat transfer enhancement in latent heat thermal

storage system for buildings. Energy and Buildings. 2003;**35**:1097-1104

[19] Alper AA, Okutan H. High-chain fatty acid esters of myristyl alcohol with odd carbon number: Novel organic phase change materials for thermal energy storage-2. Solar Energy Materials and Solar Cells. 2011;**95**(8):2417-2423

[20] Tyagi VV, Kaushik SC, Tyagi SK, Akiyama T. Development of phase change materials based microencapsulated technology for buildings: A review. Renewable and Sustainable Energy Reviews. 2011;**15**(2):1373-1391

[21] Soibam J. Numerical investigation of a heat exchanger using phase change materials [M.Sc. thesis]. Tronhiem, Norway, NTNU: Norwegian University of Science and Technology; 2017

[22] Rathod MK. Thermal stability of phase change materials. In: Phase Change Materials and Their Applications. London, UK: IntechOpen; 2018

[23] Raoux S, Wutting M. Phase Change Materials—Science and Application. Boston, MA, USA: Springer; 2009

[24] Khan Z, Khan Z, Ghafoor A. A review of performance enhancement of PCM based latent heat storage system within the context of materials, thermal stability and compatibility. Energy Conversion and Management. 2016;**115**:132-158

[25] Haynes WM. CRC Handbook of Chemistry and Physics. 91st ed. Boca Raton, FL: CRC Press Inc.; 2010-2011

[26] Hirschey J, Gluesenkamp KR, et al. Review of inorganic salt hydrates with phase change temperature in range of 5°C to 60°C and material cost comparison with common waxes. In: 5th International High Performance Buildings Conference at Purdue. USA: Predue University; July 9-12, 2018

[27] Streicher W, Cabeza L, Heinz A. Inventory of Phase Change Materials (PCM). A Report of IEA Solar Heating and Cooling Programme—Task 32, Advanced Storage Concepts for Solar and Low Energy Buildings. Austria: Graz University of Technology; 2005

[28] Zhang H. Sugar alcohol based heat storage materials: A nanoscale study and beyond [PhD thesis]. Eindhoven, Netherland: Eindhoven University of Technology; 2017

[29] Gulfam R, Zhang P, Meng Z. Advanced thermal systems driven by paraffin-based phase change materials—A review. Applied Energy. 2019;**238**:582-611

[30] Almadhoni K. A review—An optimization of macro-encapsulated paraffin used in solar latent heat storage unit. International Journal of Engineering Research and Technology. 2016;**5**(1):729-736

[31] Calvet N, Py X, et al. Enhanced performances of macro-encapsulated phase change materials (PCMs) by intensification of the internal effective thermal conductivity. Energy. 2013;**55**:956-964

[32] Teng T, Yu C. Characteristics of phase-change materials containing oxide nano-additives for thermal storage. Nanoscale Research Letters. 2012;**7**:611

[33] Karunamurthy K, Murugumohankumar K, Suresh S. Use of CuO nano-material for the improvement of thermal conductivity and performance of low temperature energy storage system of solar pond. Digest Journal of Nanomaterials and Biostructures. 2012;**7**(4):1833-1841

[34] Huang K, Liang D, et al. Macro-encapsulated PCM cylinder module

based on paraffin and float stones. Materials. 2016;**9**:361

[35] Dzhonova-Atansova DB, Georgiev AG, Popov RK. Numerical study of heat transfer in macro-encapsulated phase change material for thermal energy storage. Bulgarian Chemical Communications. 2016;**48**(Special Issue E):189-194

[36] Rahman A, Dickinson ME, Farid MM. Microencapsulation of a PCM through membrane emulsification and nanocompression-based determination of microcapsule strength. Materials for Renewable and Sustainable Energy. 2012;**1**(4)

[37] Al-shannaq R, Farid M, Dickinson M, Behzadi S. Microencapsulation of phase change materials for thermal energy storage in building application. In: Chemeca 2010: Quality of Life through Chemical Engineering: September 2012, Wellington, New Zealand. Barton, A.C.T: Engineers Australia; 2012. pp. 943-952

[38] Liu X, Lou Y. Preparation of microencapsulated phase change materials by the sol-gel process and its application on textiles. Fibres & Textiles in Eastern Europe. 2015;**23**(2(110)):63-67

[39] Li B, Liu T, Hu L, et al. Fabrication and properties of microencapsulated paraffin@SiO$_2$ phase change composite for thermal energy storage, ACS sustainable. Chemical Engineer. 2013;**1**:374-380

[40] Zhang H, Wang X, Wu D. Silica encapsulation of n-octadecane via sol–gel process: A novel microencapsulated phase-change material with enhanced thermal conductivity and performance. Journal of Colloid and Interface Science. 2010;**343**:246-255

[41] Wang LY, Tsai PS, Yang YM. Preparation of silica microspheres

encapsulating phase-change material by sol-gel method in O/W emulsion. Journal of Microencapsulation. 2006;**23**(1):3-14

[42] Sari A, Alkan C, et al. Micro/nano-encapsulated *n*-nonadecane with poly(methyl methacrylate) shell for thermal energy storage. Energy Conversion and Management. 2014;**86**:614-621

[43] Sari A, Alkan C, et al. Micro/nano-encapsulated *n*-heptadecane with polystyrene shell for latent heat thermal energy storage. Solar Energy Materials and Solar Cells. 2014;**126**:42-50

[44] Jin Z, Wang Y, Liu J, Yang Z. Synthesis and properties of paraffin capsules as phase change materials. Polymer. 2008;**49**:2903-2910

[45] Zhang X, Fan Y, Tao X, Yick K. Crystallization and prevention of supercooling of microencapsulated n-alkanes. Journal of Colloid and Interface Science. 2005;**281**:299-306

[46] Silakhori M, Metselaar HSC, Mahlia TMI, Fauzi H. Preparation and characterisation of microencapsulated paraffin wax with polyaniline-based polymer shells for thermal energy storage. Materials Research Innovations. 2014;**18**(6):480-484

[47] Mohammadi B, Seyyed Najafi F, et al. Microencapsulation of butyl palmitate in polystyrene-co-methyl methacrylate Shell for thermal energy storage application. Iranian Journal of chemistry and chemical Engineering. 2018;**37**(3):187-194

[48] Su JF, Huang Z. Fabrication and properties of microencapsulated-paraffin/gypsum-matrix building materials for thermal energy storage. Energy Conversion and Management. 2012;**55**:101-107

[49] Singh KGK et al. Micro-encapsulation of paraffin wax

microspheres with silver. Defense Science Journal. 2018;**68**(2):218-224

[50] Sanchez Silva L et al. Microencapsulation of PCMs with a styrene-methyl methacrylate copolymer shell by suspension-like polymerization. Chemical Engineering Journal. 2010;**157**:216-222

[51] Yuan K, Wang H, Zhang Z. Novel slurry containing graphene oxide-grafted microencapsulated phase change material with enhanced thermo-physical properties and photo-thermal performance. Solar Energy Materials and Solar Cells. 2015;**143**:29-37

[52] Zhang L, Yang W, et al. Graphene oxide-modified microencapsulated phase change materials with high encapsulation capacity and enhanced leakage-prevention performance. Applied Energy. 2017;**197**:354-363

[53] Sahan N, Fois M, Paksoy H. Improving thermal conductivity phase change materials—A study of paraffin nanomagnetite composites. Solar Energy Materials and Solar Cells. 2015;**137**:61-67

[54] Chaichan M, Kamel H, Al-Ajeely M. Thermal conductivity enhancement by using Nano-material in phase change material for latent heat thermal energy storage systems. SAUSSUREA. 2015;**5**:48-55

[55] Yanlai Z, Wang S, Rao Z, Xie J. Experiment on heat storage characteristic of microencapsulated phase change material slurry. Solar Energy Materials and Solar Cells. 2011;**95**(10):2726-2733

[56] Huang MJ, Eames PC, Norton B, Hewitt NJ. Natural convection in an internally finned phase change material heat sink for the thermal management of photovoltaics. Solar Energy Materials and Solar Cells. 2011;**95**(7):1598-1603

[57] Vakhshouri AR, Azizov AH, Aliyeva RV, et al. Preparation and study of thermal properties of phase change materials based on paraffin-alumina filled polyethylene. Journal of Applied Polymer Science. 2011;**120**(4):1907-1915

[58] Mu M, Basheer PAM. Shape stabilized phase change materials based on a high melt viscosity HDPE and paraffin waxes. Applied Energy. 2016;**162**:68-82

[59] Inaba H, Tu P. Evaluation of thermo-physical characteristics on shape-stabilized paraffin as a solid–liquid phase change material. Heat and Mass Transfer. 1997;**32**:307-312

[60] Feldman D, Shapiro M, Fazio P. A heat storage module with a polymer structural matrix. Polymer Engineering & Science. 1985;**25**(7):406-411

[61] Lee C, Choi HK. Crystalline morphology in high density polyethylene/paraffin blend for thermal energy storage. Polymer Composites. 1998;**19**(6):704-708

[62] Hong Y, Xin-shi G. Preparation of polyethylene–paraffin compound as a form-stable solid–liquid phase change material. Solar Energy Materials and Solar Cells. 2000;**64**:37-44

[63] Xiao M, Feng B, Gong K. Preparation and performance of shape stabilized phase change thermal storage materials with high thermal conductivity. Energy Conversion and Management. 2002;**43**:103-108

[64] Salyer IO. Phase Change Materials Incorporated throughout the Structure of Polymer Fibers. Pat. US 5885475; 1999

[65] Sari A. Form-stable paraffin/high density polyethylene composites as solid–liquid phase change materials for thermal energy storage: Preparation and

thermal properties. Energy Conversion and Management. 2004;**45**:2033-2042

[66] Zhang ZG, Fang XM. Study on paraffin/expanded graphite composite phase change thermal energy storage material. Energy Conversion and Management. 2006;**47**:303-310

[67] Zhang YP, Ding JH, Wang X, Ang R, Lin KP. Influence of additives on thermal conductivity of shape-stabilized phase change material. Solar Energy Materials & Solar Cells. 2006;**90**:1692-1702

[68] Wang J, Xie H, Lee A, Xin Z. PW based phase change nanocomposites containing γ-Al$_2$O$_3$. Journal of Thermal Analysis and Calorimetry. 2010;**102**:709-713

[69] Cai YB, Hu Y, Song L, Tang Y. Flammability and thermal properties of high density polyethylene/paraffin hybrid as a form-stable phase change material. Journal of Applied Polymer Science. 2006;**99**:1320-1327

[70] Cai YB, Hu Y, Song L, Kong QH. Preparation and flammability of high density polyethylene/paraffin/organophilic montmorillonite hybrids as a form-stable phase change material. Energy Conversion and Management. 2007;**48**:462-469

[71] Cai YB, Song L, He Q, Ang D, Hu Y. Preparation, thermal and flammability properties of a novel form-stable phase change materials based on high density polyethylene/poly (ethylene-co-vinyl acetate)/organophilic montmorillonite nanocomposites/paraffin compounds. Energy Conversion and Management. 2008;**49**:2055-2062

[72] Cai YB, Wei Q, Huang F, Gao W. Preparation and properties studies of halogen-free flame retardant form-stable phase change materials based on paraffin/high density polyethylene composites. Applied Energy. 2008;**85**:765-775

[73] Costa FR, Wagenknecht U, Heinrich G. LDPE/Mg-Al layered double hydroxide nanocomposite: Thermal and flammability properties. Polymer Degradation and Stability. 2007;**92**:1813-1823

[74] Jizhao L, Yingjie Z. A study of the flame-retardant properties of polypropylene/Al(OH)$_3$/Mg(OH)$_2$ composites. Polymer International. 2010;**59**(4):539-542

[75] Meshkova IN, Petrosen AI, Lalaen VM, Dubnikova IL. A comparative analysis of polymerization filled polyethylene composites with reduced flammability. Vysokomolekulernye Soedinenie, Series B. 2008;**50**(5):918-925

[76] Jiang Y, Yan P, et al. Form-stable phase change materials with enhanced thermal stability and fire resistance via the incorporation of phosphorus and silicon. Materials and Design. 2018;**160**:763-771

[77] Xu Y, He Y, et al. Al/Al$_2$O$_3$ form-stable phase change material for high temperature thermal energy storage. Energy Procedia. 2017;**105**:4328-4333

[78] Hasanabadi S, Sadrameli SM, et al. A cost-effective form stable PCM composite with modified paraffin and expanded perlite for thermal energy storage in concrete. Journal of Thermal Analysis and Calorimetry. 2019;**136**:1201

[79] Rawi S, Amin M, et al. Characterization of shape-stabilized phase change material using beeswax and functionalized multi-walled carbon nanotubes. IOP Conference Series: Earth and Environmental Science. 2018;**105**:012042

[80] Wan Y, Chen Y, Cui Z, et al. A promising form-stable phase change material prepared using cost effective pinecone biochar as the matrix of

palmitic acid for thermal energy storage. Scientific Reports. 2019;**9**:11535

[81] Wang Y, Xia TD, Zheng H, Feng HX. Stearic acid/silica fume composite as form-stable phase change material for thermal energy storage. Energy and Buildings. 2011;**43**(9):2365-2370

[82] Fan L, Khodadadi JM. Thermal conductivity enhancement of phase change materials for thermal energy storage: A review. Renewable and Sustainable Energy Reviews. 2011;**15**(1):24-46

[83] Mei D, Zhang B, Liu R, Zhang Y, Liu J. Preparation of capric acid/halloysite nanotube composite as form-stable phase change material for thermal energy storage. Solar Energy Materials and Solar Cells. 2011;**95**(10):2772-2777

[84] Zhang X, Deng P, Feng R, Song J. Novel gelatinous shape-stabilized phase change materials with high heat storage density. Solar Energy Materials and Solar Cells. 2011;**95**(4):1213-1218

[85] Kuznik F, David D, Johannes K, Roux J. A review on phase change materials integrated in building walls. Renewable and Sustainable Energy Reviews. 2011;**15**:379-391

[86] Elgafy A, Lafdi K. Effect of carbon nanofiber additives on thermal behavior of phase change materials. Carbon. 2005;**43**:3067-3074

[87] Rossi RM, Bolli WP. Phase change materials for improvement of heat protection. Advanced Engineering Materials. 2005;7:368-373

[88] Bremerkamp F, Seehase D, Nowottnick M. Heat protection coatings for high temperature electronics. In: 35th International Spring Seminar on Electronics Technology. Austria: Vienna University of Technology; 2012

[89] Resch-Fauster K, Hengstberger F, et al. Overheating protection of solar thermal façades with latent heat storages based on paraffin-polymer compounds. Energy and Buildings. 2018;**169**:254-259

[90] Zhu N, Ma Z, Wang S. Dynamic characteristics and energy performance of buildings using phase change materials: A review. Energy Conversion and Management. 2009;**50**:3169-3181

[91] Cui Y, Xie J, et al. A review on phase change material application in building. Advances in Mechanical Engineering. 2017;**9**(6):1-15

[92] Tyagi VV, Buddhi D. PCM thermal storage in buildings: A state of art. Renewable & Sustainable Energy Reviews. 2007;**11**:1146-1166

[93] Diaconua BM. Thermal energy savings in buildings with PCM-enhanced envelope: Influence of occupancy pattern and ventilation. Energy and Buildings. 2011;**43**:101-107

[94] Diaconu B, Cruceru M. Novel concept of composite phase change material (PCM) wall system for year-round thermal energy savings. Energy and Buildings. 2010;**42**:1759-1772

[95] Darkwa K, O'Callaghan PW. Simulation of phase change drywalls in a passive solar building. Applied Thermal Engineering. 2006;**26**